Environmental Monitoring

Environmental Monitoring

Edited by
Hayden Cole

☐ Larsen & Keller
www.larsen-keller.com

Environmental Monitoring
Edited by Hayden Cole
ISBN: 978-1-63549-112-8 (Hardback)

© 2017 Larsen & Keller

Larsen & Keller

Published by Larsen and Keller Education,
5 Penn Plaza,
19th Floor,
New York, NY 10001, USA

Cataloging-in-Publication Data

Environmental Monitoring / edited by Hayden Cole.
   p. cm.
Includes bibliographical references and index.
ISBN 978-1-63549-112-8
   1. Environmental monitoring. 2. Environmental
engineering. 3. Pollution--Measurement.
I. Cole, Hayden.
QH541.15.M64 E58 2017
628--dc23

The publisher's policy is to use permanent paper from mills that operate a sustainable forestry policy. Furthermore, the publisher ensures that the text paper and cover boards used have met acceptable environmental accreditation standards.

Printed and bound in the United States of America.

For more information regarding Larsen and Keller Education and its products, please visit the publisher's website www.larsen-keller.com

Table of Contents

Preface

This book is compiled in such a manner, that it will provide in-depth knowledge about the theory and practice of environmental monitoring. It will discuss in detail about the different applications of this field. Environmental monitoring refers to practices and activities used to monitor the quality of the environment. It is also used to compile environmental impact assessments, which examines the amount of damage human activities cause to the natural environment. The different types of monitoring include air monitoring, water quality monitoring and soil monitoring. Some of the diverse topics covered in this book address the varied branches that fall under this subject area. The topics discussed are of utmost significance and are bound to provide incredible insights to students. The textbook is appropriate for those seeking detailed information in this area.

Given below is the chapter wise description of the book:

Chapter 1- Environmental issues have alarmingly increased in the last few decades. To monitor these changes there are processes and activities and these developments are known as environmental monitoring. It helps in the preparation of disasters that are caused by anthropological activities. This chapter will provide an integrated understanding of environmental monitoring.

Chapter 2- Tools and techniques are important components of any field of study. Some of the techniques elucidated in this section are particulate matter sampler, portable emissions measurement system, contour line, gas chromatography, chemcatcher, sorbent tube etc. Particulate matter sampler is a tool, which is used to measure the properties of particulates in air whereas gas chromatography is a technique of separating and then analyzing compounds that can be vaporized. The following text elucidates the various technologies that are related to environmental monitoring.

Chapter 3- Monitoring in air pollution can be done in various ways. Governments to communicate the present population of air in a particular area use air quality index. Some other ways of environmental monitoring are air quality index, upper-atmospheric models, indoor air quality, wind rose and atmospheric dispersion modelling. This section on environmental monitoring offers an insightful focus, keeping in mind the complex subject matter.

Chapter 4- Water pollution has become one of the major problems faced by our world in today's time. Considering how vital water is to life, it deserves our utmost attention. There are a number of ways to monitor water pollution; some of these are water quality modeling, bacteriological water analysis, wastewater quality indicators and water- sensitive urban design. Water pollution can best be understood in confluence with the major topics listed in the following section.

Chapter 5- Radiation monitoring involves the measurement of radiation dose. It indicates the toxicity of radioactive substances. Some of the topics discussed in this chapter are absorbed dose, radionuclide, ionization chamber, gaseous ionization detectors and scintillation counter. This chapter will provide an integrated understanding of radiation monitoring.

Chapter 6- Environmental concerns and challenges have drastically increased in the past few years. Some of the major concerns related to the environment are drug pollution, air pollution, greenhouse gas, acid rain and chemical waste. The pollution that is caused by pharmaceutical drugs by dumping their waste in the marine environment causes drug pollution as well as marine pollution. The topics discussed in the chapter are of great importance to broaden the existing knowledge on the environment.

Chapter 7- Human activities have greatly affected the environment. This can be described by the formula I = PAT, in words this means the impact humans have on the environment equals to the production of population, affluence and technology. Human impacts on the nitrogen cycle and life cycle assessment have also been explained in the following section. This chapter is a compilation of the various branches of environmental monitoring that form an integral part of the broader subject matter.

Indeed, my job was extremely crucial and challenging as I had to ensure that every chapter is informative and structured in a student-friendly manner. I am thankful for the support provided by my family and colleagues during the completion of this book.

Editor

Introduction to Environmental Monitoring

Environmental issues have alarmingly increased in the last few decades. To monitor these changes there are processes and activities and these developments are known as environmental monitoring. It helps in the preparation of disasters that are caused by anthropological activities. This chapter will provide an integrated understanding of environmental monitoring.

Environmental monitoring describes the processes and activities that need to take place to characterise and monitor the quality of the environment. Environmental monitoring is used in the preparation of environmental impact assessments, as well as in many circumstances in which human activities carry a risk of harmful effects on the natural environment. All monitoring strategies and programmes have reasons and justifications which are often designed to establish the current status of an environment or to establish trends in environmental parameters. In all cases the results of monitoring will be reviewed, analysed statistically and published. The design of a monitoring programme must therefore have regard to the final use of the data before monitoring starts.

Air Quality Monitoring

Air quality monitoring is performed using specialized equipment and analytical methods used to establish air pollutant concentrations.

Air monitors are operated by citizens, regulatory agencies, and researchers to investigate air quality and the effects of air pollution.

Interpretation of ambient air monitoring data often involves a consideration of the spatial and temporal representativeness of the data gathered, and the health effects associated with exposure to the monitored levels.

Since air pollution is carried by the wind, consideration of anemometer data in the area between sources and the monitor often provides insights on the source of the air contaminants recorded by an air pollution monitor.

Close to the earth's surface, the atmosphere normally gets colder with height, but on certain days, the atmosphere begins to get warmer with height a short distance from the earth's surface, and air emissions build up under this "cap" on the vertical mixing.

Topographic features (such as a valley) that prevent lateral atmospheric mixing, coupled with the vertical cap on atmospheric mixing caused by an inversion, can lead to especially high air pollutant concentrations, for example, the 1948 Donora smog.

An inversion in a mountain valley in Poland

Air dispersion models that combine topographic, emissions and meteorological data to predict air pollutant concentrations are often helpful in interpreting air monitoring data.

If an air monitor produces concentrations of multiple chemical compounds, a unique "chemical fingerprint" of a particular air pollution source may emerge from analysis of the data.

Soil Monitoring

Soil monitoring is the process of collection of soil and testing in laboratory by analytical methods.

Soil sampling are of two types:

- Grab Sampling: in this method, sample is collected randomly from field

- Composite Sampling: In this method, mixing of multiple sub samples for larger and non-uniform fields.

In laboratory, soil can be tested for pH, Chlorides, Sulphates, Phosphates and other metals.

Water Quality Monitoring

Design of Environmental Monitoring Programmes

Water quality monitoring is of little use without a clear and unambiguous definition of the reasons for the monitoring and the objectives that it will satisfy. Almost all monitoring (except perhaps remote sensing) is in some part invasive of the environment under study and extensive and poorly planned monitoring carries a risk of damage to the environment. This may be a critical consideration in wilderness areas or when monitoring very rare organisms or those that are averse to human presence. Some monitoring techniques, such gill netting fish to estimate populations, can be very damaging, at least to the local population and can also degrade public trust in scientists carrying out the monitoring.

Almost all mainstream environmentalism monitoring projects form part of an overall monitoring strategy or research field, and these field and strategies are themselves derived from the high levels objectives or aspirations of an organisation. Unless individual monitoring projects fit into a wider strategic framework, the results are unlikely to be published and the environmental understanding produced by the monitoring will be lost.

Chemical

Analyzing Water Samples for Pesticides

The range of chemical parameters that have the potential to affect any ecosystem is very large and in all monitoring programmes it is necessary to target a suite of parameters based on local knowledge and past practice for an initial review. The list can be expanded or reduced based on developing knowledge and the outcome of the initial surveys.

Freshwater environments have been extensively studied for many years and there is a robust understanding of the interactions between chemistry and the environment across much of the world. However, as new materials are developed and new pressures come to bear, revisions to monitoring programmes will be required. In the last 20 years acid rain, synthetic hormone analogues, halogenated hydrocarbons, greenhouse gases and many others have required changes to monitoring strategies.

Biological

In ecological monitoring, the monitoring strategy and effort is directed at the plants and animals in the environment under review and is specific to each individual study.

However, in more generalised environmental monitoring, many animals act as robust indicators of the quality of the environment that they are experiencing or have experienced in the recent past. One of the most familiar examples is the monitoring of numbers of Salmonid fish such as Brown trout or Salmon in river systems and lakes to detect slow trends in adverse environmental effects. The steep decline in salmonid fish populations was one of the early indications of the problem that later became known as acid rain.

In recent years much more attention has been given to a more holistic approach in which the ecosystem health is assessed and used as the monitoring tool itself. It is this approach that underpins the monitoring protocols of the Water Framework Directive in the European Union.

Radiological

Radiation monitoring involves the measurement of radiation dose or radionuclide contamination for reasons related to the assessment or control of exposure to ionizing radiation or radioactive substances, and the interpretation of the results. The 'measurement' of dose often means the measurement of a dose equivalent quantity as a proxy (i.e. substitute) for a dose quantity that cannot be measured directly. Also, sampling may be involved as a preliminary step to measurement of the content of radionuclides in environmental media. The methodological and technical details of the design and operation of monitoring programmes and systems for different radionuclides, environmental media and types of facility are given in IAEA Safety Guide RS–G-1.8 and in IAEA Safety Report No. 64.

Radiation monitoring is often carried out using networks of fixed and deployable sensors such as the US Environmental Protection Agency's Radnet and the SPEEDI network in Japan. Airborne surveys are also made by organizations like the Nuclear Emergency Support Team.

Microbiological

Bacteria and viruses are the most commonly monitored groups of microbiological organisms and even these are only of great relevance where water in the aquatic environment is subsequently used as drinking water or where water contact recreation such as swimming or canoeing is practised.

Although pathogens are the primary focus of attention, the principal monitoring effort is almost always directed at much more common indicator species such as *Escherichia coli,* supplemented by overall coliform bacteria counts. The rationale behind this monitoring strategy is that most human pathogens originate from other humans via the sewage stream. Many sewage treatment plants have no sterilisation final stage and therefore discharge an effluent which, although having a clean appearance, still contains many millions of bacteria per litre, the majority of which are relatively harmless coliform bacteria. Counting the number of harmless (or less harmful) sewage bacteria allows a judgement to be made about the probability of significant numbers of pathogenic bacteria or viruses being present. Where *E. coli* or coliform levels exceed pre-set trigger values, more intensive monitoring including specific monitoring for pathogenic species is then initiated.

Populations

Monitoring strategies can produce misleading answers when relaying on counts of species or presence or absence of particular organisms if there is no regard to population size. Understanding the populations dynamics of an organism being monitored is critical.

As an example if presence or absence of a particular organism within a 10 km square is the measure adopted by a monitoring strategy, then a reduction of population from 10,000 per square to 10 per square will go unnoticed despite the very significant impact experienced by the organism.

Monitoring Programmes

All scientifically reliable environmental monitoring is performed in line with a published programme. The programme may include the overall objectives of the organisation, references to the specific strategies that helps deliver the objective and details of specific projects or tasks within those strategies. However the key feature of any programme is the listing of what is being monitored and how that monitoring is to take place and the time-scale over which it should all happen. Typically, and often as an appendix, a monitoring programme will provide a table of locations, dates and sampling methods that are proposed and which, if undertaken in full, will deliver the published monitoring programme.

There are a number of commercial software packages which can assist with the implementation of the programme, monitor its progress and flag up inconsistencies or omissions but none of these can provide the key building block which is the programme itself.

Environmental Monitoring Data Management Systems

Given the multiple types and increasing volumes and importance of monitoring data, commercial software Environmental Data Management Systems (EDMS) or E-MDMS are increasingly in common use by regulated industries. They provide a means of managing all monitoring data in a single central place. Quality validation, compliance checking, verifying all data has been received, and sending alerts are generally automated. Typical interrogation functionality enables comparison of data sets both temporarily and spatially. They will also generate regulatory and other reports.

Formal Certification:

(May 2014) there is only one certification scheme specifically for environmental data management software. This is provided by the Environment Agency in the UK under its Monitoring Certification Scheme (MCERTS).

Sampling Methods

There are a wide range of sampling methods which depend on the type of environment, the material being sampled and the subsequent analysis of the sample.

At its simplest a sample can be filling a clean bottle with river water and submitting it

for conventional chemical analysis. At the more complex end, sample data may be produced by complex electronic sensing devices taking sub-samples over fixed or variable
time periods.

Grab Samples

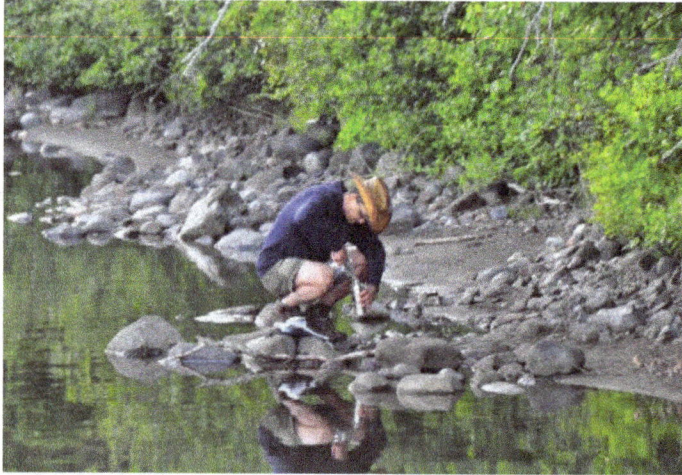

Collecting a grab sample on a stream

Grab samples are samples taken of a homogeneous material, usually water, in a single
vessel. Filling a clean bottle with river water is a very common example. Grab samples
provide a good snap-shot view of the quality of the sampled environment at the point of
sampling and at the time of sampling. Without additional monitoring, the results cannot be extrapolated to other times or to other parts of the river, lake or ground-water.

In order to enable grab samples or rivers to be treated as representative, repeat trans-
verse and longitudinal transect surveys taken at different times of day and times of year
are required to establish that the grab-sample location is as representative as is reasonably possible. For large rivers such surveys should also have regard to the depth of the
sample and how to best manage the sampling locations at times of flood and drought.

In lakes grab samples are relatively simple to take using depth samplers which can be
lowered to a pre-determined depth and then closed trapping a fixed volume of water
from the required depth. In all but the shallowest lakes, there are major changes in the
chemical composition of lake water at different depths, especially during the summer
months when many lakes stratify into a warm, well oxygenated upper layer (*epilimni-
on*) and a cool de-oxygenated lower layer *(hypolimnion)*.

In the open seas marine environment grab samples can establish a wide range of baseline parameters such as salinity and a range of cation and anion concentrations. However, where changing conditions are an issue such as near river or sewage discharges,
close to the effects of volcanism or close to areas of freshwater input from melting ice, a
grab sample can only give a very partial answer when taken on its own.

Semi-continuous Monitoring and Continuous

An automated sampling station and data logger (to record temperature, specific conductance, and dissolved oxygen levels)

There is a wide range of specialized sampling equipment available that can be programmed to take samples at fixed or variable time intervals or in response to an external trigger. For example, a sampler can be programmed to start taking samples of a river at 8 minute intervals when the rainfall intensity rises above 1 mm / hour. The trigger in this case may be a remote rain gauge communicating with the sampler by using cell phone or meteor burst technology. Samplers can also take individual discrete samples at each sampling occasion or bulk up samples into composite so that in the course of one day, such a sampler might produce 12 composite samples each composed of 6 sub-samples taken at 20 minute intervals.

Continuous or quasi-continuous monitoring involves having an automated analytical facility close to the environment being monitored so that results can, if required, be viewed in real time. Such systems are often established to protect important water supplies such as in the River Dee regulation system but may also be part of an overall monitoring strategy on large strategic rivers where early warning of potential problems is essential. Such systems routinely provide data on parameters such as pH, dissolved oxygen, conductivity, turbidity and colour but it is also possible to operate gas liquid chromatography with mass spectrometry technologies (GLC/MS) to examine a wide range of potential organic pollutants. In all examples of automated bank-side analysis there is a requirement for water to be pumped from the river into the monitoring station. Choosing a location for the pump inlet is equally as critical as deciding on the location for a river grab sample. The design of the pump and pipework also requires careful design to avoid artefacts being introduced through the action of pumping the water. Dissolved oxygen concentration is difficult to sustain through a pumped system and GLC/MS facilities can detect micro-organic contaminants from the pipework and glands.

Passive Sampling

The use of passive samplers greatly reduces the cost and the need of infrastructure on the sampling location. Passive samplers are semi-disposable and can be produced at a relatively low cost, thus they can be employed in great numbers, allowing for a better cover and more data being collected. Due to being small the passive sampler can also be hidden, and thereby lower the risk of vandalism. Examples of passive sampling devices are the diffusive gradients in thin films (DGT) sampler, Chemcatcher, Polar organic chemical integrative sampler (POCIS), and an air sampling pump.

Remote Surveillance

Although on-site data collection using electronic measuring equipment is common-place, many monitoring programmes also use remote surveillance and remote access to data in real time. This requires the on-site monitoring equipment to be connected to a base station via either a telemetry network,land-line, cell phone network or other telemetry system such as Meteor burst. The advantage of remote surveillance is that many data feeds can come into a single base station for storing and analysis. It also enable trigger levels or alert levels to be set for individual monitoring sites and/or parameters so that immediate action can be initiated if a trigger level is exceeded. The use of remote surveillance also allows for the installation of very discrete monitoring equipment which can often be buried, camouflaged or tethered at depth in a lake or river with only a short whip aerial protruding. Use of such equipment tends to reduce vandalism and theft when monitoring in locations easily accessible by the public.

Remote Sensing

Environmental remote sensing uses aircraft or satellites to monitor the environment using multi-channel sensors.

There are two kinds of remote sensing. Passive sensors detect natural radiation that is emitted or reflected by the object or surrounding area being observed. Reflected sunlight is the most common source of radiation measured by passive sensors and in environmental remote sensing, the sensors used are tuned to specific wavelengths from far infra-red through visible light frequencies through to far ultra violet. The volumes of data that can be collected are very large and require dedicated computational support . The output of data analysis from remote sensing are false colour images which differentiate small differences in the radiation characteristics of the environment being monitored. With a skilful operator choosing specific channels it is possible to amplify differences which are imperceptible to the human eye. In particular it is possible to discriminate subtle changes in chlorophyll a and chlorophyll b concentrations in plants and show areas of an environment with slightly different nutrient regimes.

Active remote sensing emits energy and uses a passive sensor to detect and measure the radiation that is reflected or backscattered from the target. LIDAR is often used to acquire information about the topography of an area, especially when the area is large and manual surveying would be prohibitively expensive or difficult.

Remote sensing makes it possible to collect data on dangerous or inaccessible areas. Remote sensing applications include monitoring deforestation in areas such as the Amazon Basin, the effects of climate change on glaciers and Arctic and Antarctic regions, and depth sounding of coastal and ocean depths.

Orbital platforms collect and transmit data from different parts of the electromagnetic spectrum, which in conjunction with larger scale aerial or ground-based sensing and analysis, provides information to monitor trends such as El Niño and other natural long and short term phenomena. Other uses include different areas of the earth sciences such as natural resource management, land use planning and conservation.

Bio-monitoring

The use of living organisms as monitoring tools has many advantages. Organisms living in the environment under study are constantly exposed to the physical, biological and chemical influences of that environment. Organisms that have a tendency to accumulate chemical species can often accumulate significant quantities of material from very low concentrations in the environment. Mosses have been used by many investigators to monitor heavy metal concentrations because of their tendency to selectively adsorb heavy metals.

Similarly, eels have been used to study halogenated organic chemicals, as these are adsorbed into the fatty deposits within the eel.

Other Sampling Methods

Ecological sampling requires careful planning to be representative and as noninvasive as possible. For grasslands and other low growing habitats the use of a quadrat – a 1-metre square frame – is often used with the numbers and types of organisms growing within each quadrat area counted

Sediments and soils require specialist sampling tools to ensure that the material recovered is representative. Such samplers are frequently designed to recover a specified volume of material and may also be designed to recover the sediment or soil living biota as well such as the Ekman grab sampler.

Data Interpretations

The interpretation of environmental data produced from a well designed monitoring programme is a large and complex topic addressed by many publications. Regrettably it is sometimes the case that scientists approach the analysis of results with a pre-conceived outcome in mind and use or misuse statistics to demonstrate that their own particular point of view is correct.

Statistics remains a tool that is equally easy to use or to misuse to demonstrate the lessons learnt from environmental monitoring.

Environmental Quality Indices

Since the start of science based environmental monitoring, a number of quality indices have been devised to help classify and clarify the meaning of the considerable volumes of data involved. Stating that a river stretch is in "Class B" is likely to be much more informative than stating that this river stretch has a mean BOD of 4.2, a mean dissolved oxygen of 85%, etc. In the UK the Environment Agency formally employed a system called General Quality Assessment (GQA) which classified rivers into six quality letter bands from A to F based on chemical criteria and on biological criteria. The Environment Agency and its devolved partners in Wales (Countryside Council for Wales, CCW) and Scotland (Scottish Environmental Protection Agency, SEPA) now employ a system of biological, chemical and physical classification for rivers and lakes that corresponds with the EU Water Framework Directive.

References

- Hart, C.W.; Fuller, Samuel F.J. (1974). Pollution Ecology of Freshwater Invertebrates. New York: Academic Press. ISBN 0-12-328450-3.

- International Atomic Energy Agency (2007). IAEA Safety Glossary: Terminology Used in Nuclear Safety and Radiation Protection (PDF). Vienna: IAEA. ISBN 92-0-100707-8.

- International Atomic Energy Agency (2010). Programmes and Systems for Source and Environmental Radiation Monitoring. Safety Reports Series No. 64. Vienna: IAEA. p. 234. ISBN 978-92-0-112409-8.

2

Techniques and Technologies of Environmental Monitoring

Tools and techniques are important components of any field of study. Some of the techniques elucidated in this section are particulate matter sampler, portable emissions measurement system, contour line, gas chromatography, chemcatcher, sorbent tube etc. Particulate matter sampler is a tool, which is used to measure the properties of particulates in air whereas gas chromatography is a technique of separating and then analyzing compounds that can be vaporized. The following text elucidates the various technologies that are related to environmental monitoring.

Particulate Matter Sampler

A particulate matter sampler is an instrument for measuring the properties (such as mass concentration or chemical composition) of particulates in the ambient air.

Types

Two different types of particulate matter samplers exist that measure particulate mass concentration: manual samplers and automated samplers..

Manual

Manual samplers draw a known volume of air through a filter. The filter is weighed on an analytical balance before and after sampling, and the difference in weight divided by the volume of air pulled through the filter gives the mass concentration of the particulate.

Automated

Automated samplers do the weighing in the field. There are two types of automated samplers in common usage: samplers that use a beta gauge for mass measurement and samplers that use a tapered element oscillating microbalance (TEOM) for mass measurement.

Beta Gauge

Beta gauge particulate samplers have an appearance that is similar to a reel to reel tape recorder. Air is pulled through a filter tape to accumulate a sample, the mass of the tape before and after sampling is determined by advancing the tape spot into the beta attenuation cell.

Tapered Element Oscillating Microbalance (TEOM)

The TEOM particulate sampler operates by drawing air through a filter attached at the tip of a glass tube. An electrical circuit places the tube into oscillation, and the resonant frequency of the tube is proportional to the square root of the mass on the filter.

Noise Based Particle Counters

Recently, microphone based instruments have been devised that monitor noise levels in specific frequency bands to predict local PNC levels. Prototypes of such instruments have been tested in Europe and in Bangalore.

Use of Inertial Separators

Particles of different sizes have different health effects. Inertial separators are used to eliminate particles outside of the desired size range. If a gas stream containing particles of different sizes is forced to turn a sharp corner, the inertia of the large particles causes them to separate from the gas stream lines. The larger particles can be collected and removed from the gas stream after collisions with the walls of the vessel.

The two common types of inertial separators are cyclones, which spin the gas stream, causing collisions of the heavier particles with the outside of the cyclone wall, and impactors, where the gas particle stream is directed at a greased metal plate and turned at the last moment, causing the larger particles to stick to the greased plate.

Modern particulate samplers use a volumetric flow control system that pulls air through the particle separator at the velocity required to achieve the desired cutpoint.

For air pollution applications, the definition of "particulate" does not include uncombined water, and water from a particulate sample must be removed before it is weighed. This can be done either by heating the sample to evaporate the water or by placing the sample in a low humidity environment before weighing.

Portable Emissions Measurement System

A portable emissions measurement system (PEMS) is essentially a lightweight 'laboratory' that is used to test and/or assess mobile source emissions (i.e. cars, trucks, buses, construction equipment, generators, trains, cranes, etc.) for the purposes of compliance, regulation, or decision-making. Early examples of vehicle emissions equipment were developed and marketed by Warren Spring Laboratory UK during the early 1990s. This equipment was used to measure on-road emissions as part of the UK Environment Research Programme. Governmental entities like United States Environmental Protection Agency (USEPA), European Union, as well as various states and private sector entities have begun to utilize PEMS in order to reduce both the costs and time involved in making mobile emissions decisions. Various state, federal, and international agencies began referring to this shorthand term in the early 2000s, and the nickname became part of industry parlance.

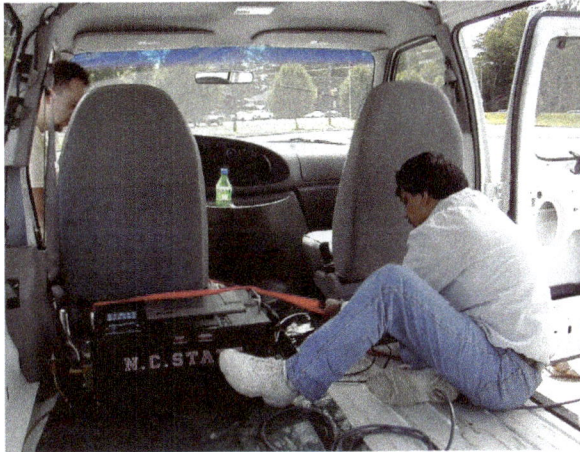

A CATI PEMS being strapped down inside a vehicle

Background

Since the mid-19th century, dynamometers (or "dyno" for short) have been used to measure torque and rotational speed (measured in rpm) from which power produced by an engine, motor or other rotating prime mover can then be calculated. A chassis dynamometer measures power from the engine through the wheels. The vehicle is parked on rollers which the car then turns and the output is measured. These dynos can be fixed or portable. Because of frictional and mechanical losses in the various drivetrain components, the measured horsepower is generally 15–20 percent less than the brake horsepower measured at the crankshaft or flywheel on an engine dynamometer. Historically though, dynamometer emission tests are very expensive, and have usually involved removing fleet vehicles from service for a long period of time. Also, the data derived from such testing is not representative of "real world" driving conditions, not least due to numerous and ample flexibilities in laboratory test procedures, particularly in the European Union's "New European Driving Dycle (NEDC)". Beyond that, this test method cannot be deemed as representative, especially due to the relatively low amount of repeatable tests at such a facility.

Introduction of PEMS

Leo Breton of the USEPA questioned dynamometer testing's ability to reflected real-world emissions levels of vehicles on-road. To investigate the matter, in 1995 he constructed a device called ROVER, short for Real-time On-road Vehicle Emissions Reporter. The first commercially available device was invented and patented by Michal Vojtisek-Lom, in conjunction with Clean Air Technologies International (CATI) Inc. out of Buffalo, New York in 1999. The early commercially available units relied upon engine data from either an "On-Board Diagnostic" (OBD) port or directly from an engine sensor array. The first unit was developed for, and sold to - Dr. H. Christopher Frey of North Carolina State University NCSU) for the first ever on-road testing project sponsored by the North Carolina Department of Transportation. David W. Miller, who co-founded CATI, first coined the phrase "Portable Emissions Measurement System" and "PEMS" when working on a 2000

A CATI PEMS testing in the field at the World Trade Center in 2002

New York City Metropolitan Transportation Agency bus project with Dr. Thomas Lanni of the New York State Department of Environmental Conservation, as a short-hand description of the new device. Other governmental groups and universities soon followed, and quickly began to utilize the equipment due to it's balance of accuracy, low cost, light weight, and availability. From 1999 through 2004, research groups such as Virginia Tech, Penn State, andTexas A&M Transportation Institute, Texas Southern University and others began to use PEMS in border crossing projects, roadway evaluations, traffic control methods, before-and-after scenarios, and even on ferries, planes, and off-road vehicles, to explore what was possible outside of a lab environment. A project performed in April of 2002 by the California Air Resources Board(ARB) - utilizing non-1065 PEMS equipment, tested 40 trucks over a period of 2 ½ days; of which, 22 trucks were tested on road in Tulare, California. Also during this timeframe, one such high-profile project performed with early PEMS equipment was the World Trade Center (WTC) Ground Zero Project in lower Manhattan, testing concrete pumpers, bulldozers, graders, and later - diesel cranes on Building #7 - 40 stories high. Other early PEMS projects such as Dr. Chris Frey's field work was used by the USEPA in the development of the MOVES Model. However, users such as regulators and vehicle manufacturers had to wait for ROVER to be commercialized to conduct actual measurements of mass emissions rather than depend on estimates of mass emissions using data the OBD port, or a direct engine measurement, in order to have a more defensible data set. This push led to a new 2005 standard known as CFR 40 Part 1065.

Many governmental entities (such as the USEPA and the United Nations Framework Convention on Climate Change or UNFCCC) have identified target mobile-source pollutants in various mobile standards as CO_2, NOx, Particulate Matter (PM), Carbon Monoxide (CO), Hydrocarbons(HC), to ensure that emissions standards are being met. Further, these governing bodies have begun adopting in-use testing program for non-road diesel engines, as well as other types of internal combustion engines, and are requiring the use of PEMS testing. It is important to delineate the various classifications of the latest 'transferable' emissions testing equipment from-time PEMS equipment, in order to best understand the desire of portability in field-testing of emissions.

Defining Portability

An important step in the evaluation of a PEMS device is to define what a PEMS device is as well as to understand various classifications of 'transferable vehicle testing equipment':

Definition of the Term "Portable"

The word portable typically conveys an object that is "Carried or moved with ease, such as a light or small typewriter." The implication is that this is a special attribution for what would otherwise generally be non-portable objects. PEMS is in contrast with lab based systems that measure in circumstances other than normal use.

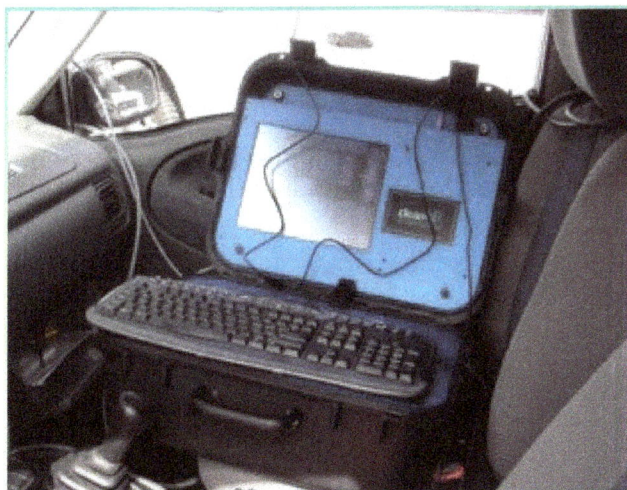

A PEMS being used to test a gasoline vehicle - 2002

Definition of the Term "Mobile"

The definition of mobile is essentially "...capable of moving or of being moved readily from place to place: a mobile organism; a mobile missile system."

Definition of "Instrumented"

Instrumented means to be "a device for recording, measuring, or controlling, especially such a device functioning as part of a control system."

Therefore, the subtle difference between 'portable' and 'mobile' is that a portable system is a lightweight device able to be carried, whereas a mobile system can be readily moved, and 'instrumented' means that the testing equipment has been incorporated into the host system. These distinctions are critical, especially considering additional guidelines from various US and International standards.

Definition Determined by the National Institute for Occupational Safety and Health (NIOSH)

The National Institute for Occupational Safety and Health (NIOSH) defines these terms based on an equation known as the "NIOSH Lifting Equation" and the "NIOSH Procedures for Analyzing Lift-ing Jobs. These clearly outline safety procedures and equipment. (these are also specified in the "Occupational Safety Hazard Act 29 CFR parts 1903, 1904, and 1910)

Safety Guidelines and Standards (the NIOSH Lifting Equation)

It is imperative to refer to existing federal standards and guidelines when determining a proper ergonomically safe and correct procedure. Not only is this important to ensure the safety of the worker(s), but also to ensure the reduction in potential future liability. Therefore, the revised NIOSH Lifting Equation is an excellent source of information to determine what a single worker should or shouldn't perform.

Based upon the NIOSH lifting equation and assuming that this diagram is analogous to the lifting of a PEMS into the cab of a heavy-duty truck the upper threshold of the total weight of a PEMS device typically should not exceed 45 lb (20 kg)., in order to be congruent with national and international safety standards. This not only allows for much more safe maneuverability and ease of use, but it also reduces the amount of workers required to safely perform such tasks.

Economic Advantage of PEMS Equipment

Because a PEMS unit is able to be carried easily by one person from jobsite to jobsite, and can be used without the requirement of 'team lifting', the required emissions testing projects are economically viable. Simply put, more testing can be done more quickly, by less workers, dramatically increasing the amount of testing done in a certain time period. This in turn, significantly reduces the 'cost per test', yet at the same time increases the overall accuracy required in a 'real-world' environment. Because the law of large numbers will create a convergence of results, it means that repeatability, predictability, and accuracy are enhanced, while simultaneously reducing the overall cost of the testing.

A next generation "integrated PEMS" (iPEMS) device.

On-road Emissions Patterns Identified by PEMS

Nearly all modern engines, when tested new and according to the accepted testing protocols in a laboratory, produce relatively low emissions well within the set standards. As all individual engines of the same series are supposed to be identical, only one or several engines of each series get tested. The tests have shown that:

1. The bulk of the total emissions can come from relatively short high-emissions episodes

2. Emissions characteristics can be different even among otherwise identical engines

3. Emissions outside of the bounds of the laboratory test procedures are often higher than under the operating and ambient conditions comparable to those during laboratory testing

4. Emissions deteriorate significantly over the useful life of the vehicles

5. There are large variances among the deterioration rates, with the high emissions rates often attributable to various mechanical malfunctions

These findings are consistent with published literature, and with the data from a myriad of subsequent studies. They are more applicable to spark-ignition engines and considerably less to diesels, but with the regulation-driven advances in diesel engine technology (comparable to the advances in spark-ignition engines since the 1970s) it can be expected that these findings are likely to be applicable to the new generation diesel engines. Since 2000, multiple entities have utilized PEMS data to measured in-use, on-road emissions on hundreds of diesel engines installed in school buses, transit buses, delivery trucks, plow trucks, over-the-road trucks, pickups, vans, forklifts, excavators, generators, loaders, compressors, locomotives, passenger ferries, and other on-road, off-road and non-road applications. All the previously listed findings were demonstrated; in addition, it was noticed that extended idling of engines can have a significant impact on the emissions during subsequent operation.

Also, PEMS testing identified several engine "anomalies" where fuel-specific NOx emissions were two to three times higher than expected during some modes of operation, suggesting deliberate alterations of the engine control unit (ECU) settings. Such data set can be readily used for developing emissions inventories, as well as to evaluate various improvements in engines, fuels, exhaust after-treatment and other areas. (Data collected on "conventional" fleets then serves as "baseline" data to which various improvements are compared.) This data set can also be examined for compliance with not-to-exceed (NTE) and in-use emissions standards, which are 'US-based' emission standards that require on-road testing.

Accuracy of PEMS

1065 PEMS manufactured by AVL - attached on a passenger car

The question often arises as to the target accuracy of PEMS. As PEMS are typically limited in size, weight and power consumption, it is often difficult for PEMS to offer the same accuracy and variety of species measured as is possible with top of the line laboratory instrumentation. For this reason, objections were raised against using PEMS for compliance verification.

On the other hand, fleet emissions deduced from laboratory measurements can be subject to significant inaccuracies if the selected engines and operating conditions were not representative of the fleet, or if deliberate anomalies (i.e., dual mapping of the ECU) were not demonstrated during laboratory testing.

The question of how accurate a monitoring system needs to be therefore cannot be objectively answered, neither can a monitoring system be easily designed, without first considering the intended application of the system and the errors associated with different approaches.

It is expected that a variety of on-board systems will be designed, ranging from suitcase-sized PEMS to instrumented trailers towed behind the tested truck. The benefits of each approach need to be considered in light of other sources of errors associated with emissions monitoring, notably vehicle-to-vehicle differences, and the emissions variability within the vehicle itself. In other words, one needs to consider the total of:

1. The difference between what is measured and what is actually emitted during a test

2. The difference between what is emitted during the test and what the vehicle emits during its everyday duties

3. The difference between the emissions characteristics of the tested vehicle and the overall emissions levels of the entire fleet.

For example, when evaluating a benefit of cleaner fuels on a fleet of city buses, one needs to compare taking a bus out of service, installing a laboratory-grade monitoring system, loading it with sandbags and driving it on a simulated route against testing several buses on their regular routes, with passengers on board, using a simpler (and possibly less accurate) monitoring system.

Additional PEMS Criteria

Another important aspect that needs to be evaluated is the safety of using PEMS on public roadways. Extensions of the tailpipe, lines and cables extending far beyond the vehicle sides, lead-acid batteries located in the passenger compartment of a bus, sharp objects, hot components accessible to bystanders, equipment blocking emergency exits or interfering with the driver, loose components likely to be caught on moving parts, and other potential hazards need to be examined. Also, any modifications to or disassembly of the tested vehicle (i.e., drilling into the exhaust, removing intake air system) need to be examined for their acceptance by both fleet managers and drivers, especially on passenger-carrying vehicles. The source of power for PEMS is a concern, as only a limited amount of power can be extracted from the vehicle electrical system. Sealed lead-acid batteries, fuel cells and generators have been used as external power sources, each with a potential significant hazard when driven on the road.

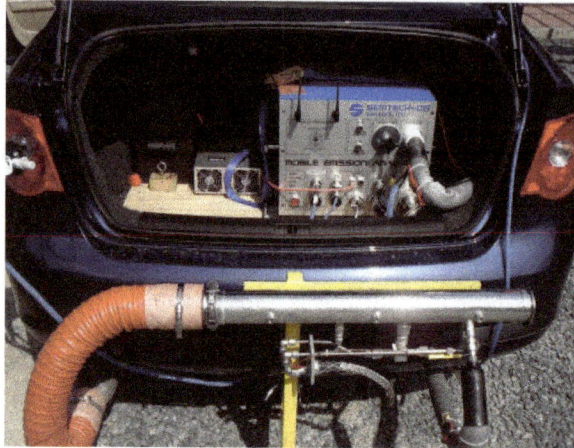

Sensors Inc. PEMS Equipment

A PEMS also has to be practical. Installation time and the expertise level required to perform installation and to operate the PEMS will have a significant impact on the cost of the test, and on the number of vehicles tested. Versatility (ability to test different vehicles) may be important if testing dissimilar engines or vehicles. Total size, weight and transportability of the PEMS needs to be considered when testing at different locations, including any consumables such as calibration gases. Any restrictions on transport of hazardous materials (i.e.Flame ionization detector (FID) fuel or calibration gases) need to be taken into the account. The ability of the test crew to repair PEMS in the field using locally available resources can also be essential when testing away from the base. Thus, PEMS evaluation protocol should be expanded. In addition to the laboratory comparison testing, which is a measure of how accurately PEMS measures when operated in a laboratory, the accuracy and repeatability of PEMS should also be examined on the road, possibly while driving along a well-defined, repeatable route, or while driving chassis dynamometer cycles on a test track.

PEMS Suitability to Application

Ultimately, it should be demonstrated to show that a PEMS is suitable to the desired application. If the ultimate goal is to verify the compliance with in-use emissions requirements, a fleet of vehicles with known characteristics – including engines with dual-mapping and otherwise non-compliant engines – should be made available for testing. It should be then up to the PEMS manufacturers to practically demonstrate how these non-compliant vehicles can be identified using their system.

Testing Volume and Safe Repeatability

In order to achieve the required amount of 'testing volume' needed to validate real-world testing, three points must be considered:

1. System accuracy

2. Federal and/or state health and safety guidelines and/or standards

3. Economic viability based on the first two points.

Once a particular portable emissions system has been identified and pronounced as accurate, the next step is to ensure that the worker(s) are properly protected from work hazards associated with the task(s) being performed in the use of the testing equipment. For example, typical functions for a worker may be to transport the equipment to the jobsite (i.e. car, truck, train, or plane), carry the equipment to the jobsite, and lift the equipment into position.

Advantages of PEMS

Next generation "integrated" PEMS equipment

On-road vehicle emissions testing is very different from the laboratory testing, bringing both considerable benefits and challenges: As the testing can take place during the regular operation of the tested vehicles, a large number of vehicles can be tested within a relatively short period of time and at relatively low cost. Engines than cannot be easily tested otherwise (i.e., ferry boat propulsion engines) can be tested. True real-world emissions data can be obtained. The instruments have to be small, lightweight, withstand difficult environment, and must not pose a safety hazard. Emissions data is subject to considerable variances, as real-world conditions are often neither well defined nor repeatable, and significant variances in emissions can exist even among otherwise identical engines. On-road emissions testing therefore requires a different mindset than the traditional approach of testing in the laboratory and using models to predict real-world performance. In the absence of established methods, use of PEMS requires careful, thoughtful, broad approach. This should be considered when designing, evaluating and selecting PEMS for the desired application.

A recent example of PEMS advantages over laboratory testing is the Volkswagen (VW) Scandal of 2015. Under a small grant from the International Council for Clean Transportation Dan Carder and Arvind Thiruvengadam of West Virginia University (WVU) uncovered on-board software "cheats" that VW had installed on some diesel passenger vehicles. The only way the discovery could have been made was by a non-programmed, random, on-road evaluation - utilizing a PEMS device. VW is now liable for over 14 billion USD in fines. In 2016, these latest developments led to a global resurgence of interest in smaller, lighter, and cost-effective "non-1065" PEMS, similar to the demonstration on the Mythbusters 2011 Premiere Episode of "Bikes and Bazookas", in which a non-1065 PEMS was used to establish the difference between car and motorcycle pollution.

Contour Line

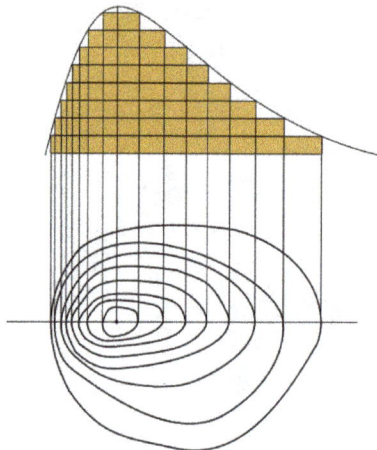

The bottom part of the diagram shows some contour lines with a straight line running through the location of the maximum value. The curve at the top represents the values along that straight line.

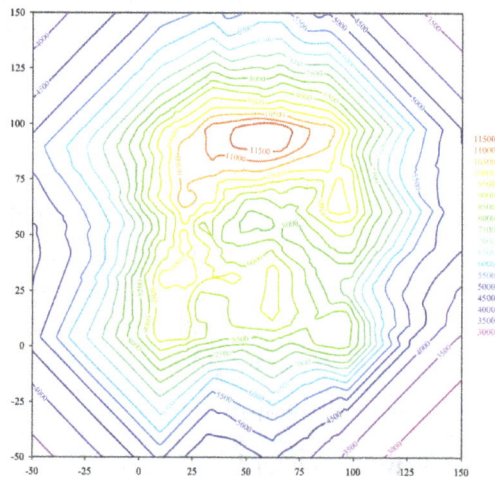

A two-dimensional contour graph of the three-dimensional surface in the above picture.

A contour line (also isoline, isopleth, or isarithm) of a function of two variables is a curve along which the function has a constant value. It is a cross-section of the three-dimensional graph of the function $f(x, y)$ parallel to the x, y plane. In cartography, a contour line (often just called a "contour") joins points of equal elevation (height) above a given level, such as mean sea level. A contour map is a map illustrated with contour lines, for example a topographic map, which thus shows valleys and hills, and the steepness of slopes. The contour interval of a contour map is the difference in elevation between successive contour lines.

More generally, a contour line for a function of two variables is a curve connecting points where the function has the same particular value. The gradient of the function is always perpendicular to the contour lines. When the lines are close together the magnitude of the gradient is large: the variation is steep. A level set is a generalization of a contour line for functions of any number of variables.

Contour lines are curved, straight or a mixture of both lines on a map describing the intersection of a real or hypothetical surface with one or more horizontal planes. The configuration of these contours allows map readers to infer relative gradient of a parameter and estimate that parameter at specific places. Contour lines may be either traced on a visible three-dimensional model of the surface, as when a photogrammetrist viewing a stereo-model plots elevation contours, or interpolated from estimated surface elevations, as when a computer program threads contours through a network of observation points of area centroids. In the latter case, the method of interpolation affects the reliability of individual isolines and their portrayal of slope, pits and peaks.

Types

Contour lines are often given specific names beginning "iso-" according to the nature of the variable being mapped, although in many usages the phrase "contour line" is most commonly used. Specific names are most common in meteorology, where multiple maps with different variables may be viewed simultaneously. The prefix "iso-" can be replaced with "isallo-" to specify a contour line connecting points where a variable changes at the same *rate* during a given time period.

The words *isoline* and *isarithm* are general terms cover-ing all types of contour line. The word *isogram* was proposed by Francis Galton in 1889 as a convenient generic designation for lines indicating equality of some physical condition or quantity; but it commonly refers to a word without a repeated letter.

An isogon is a contour line for a variable which measures direction. In meteorology and in geomagnetics, the term *isogon* has specific meanings which are described below. An isocline is a line joining points with equal slope. In population dynamics and in geomagnetics, the terms *isocline* and *isoclinic line* have specific meanings which are described below.

Equidistants (Isodistances)

Equidistant is a line of equal distance from a given point, line, polyline.

Isopleths

In geography, the word *isopleth* is used for contour lines that depict a variable which cannot be measured at a point, but which instead must be calculated from data collected over an area. An example is population density, which can be calculated by dividing the population of a census district by the surface area of that district. Each calculated value is presumed to be the value of the variable at the centre of the area, and isopleths can then be drawn by a process of interpolation. The idea of an isopleth map can be compared with that of a choropleth map.

In meteorology, the word *isopleth* is used for any type of contour line.

Meteorology

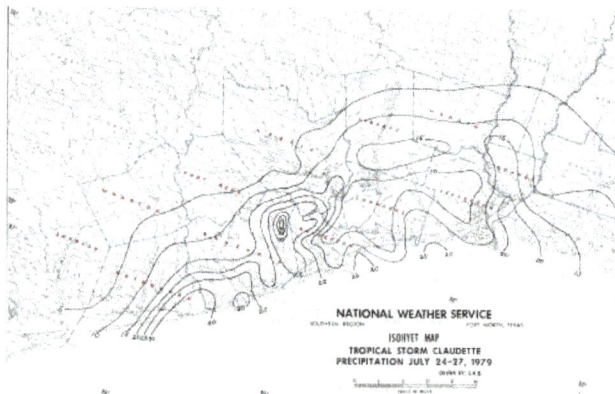

Isohyetal map of precipitation

Meteorological contour lines are based on generalization from the point data received from weather stations. Weather stations are seldom exactly positioned at a contour line (when they are, this indicates a measurement precisely equal to the value of the contour). Instead, lines are drawn to best approximate the locations of exact values, based on the scattered information points available.

Meteorological contour maps may present collected data such as actual air pressure at a given time, or generalized data such as average pressure over a period of time, or forecast data such as predicted air pressure at some point in the future

Thermodynamic diagrams use multiple overlapping contour sets (including isobars and isotherms) to present a picture of the major thermodynamic factors in a weather system.

Barometric Pressure

An isobar is a line of equal or constant pres-sure on a graph, plot, or map; an isopleth or contour line of pressure. More accurately, isobars are lines drawn on a map joining places of equal average atmospheric pressure reduced to sea level for a specified period of time. In meteorology, the barometric pres-sures shown are reduced to sea level, not the surface pressures at the map locations. The distribution of isobars is closely related to the magnitude and direction of the wind field, and can be used to predict future weather patterns. Isobars are commonly used in television weather reporting.

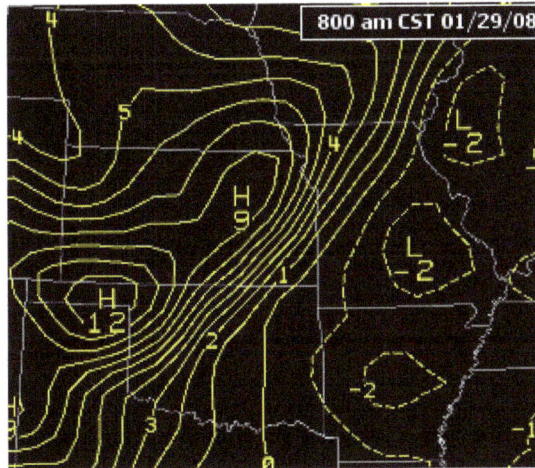

Loop showing the motion of a cold front by the movement of isallobars

Isallobars are lines joining points of equal pressure change during a specific time in-terval. These can be divided into *anallobars*, lines joining points of equal pressure in-crease during a specific time interval, and *katallobars*, lines joining points of equal pressure decrease. In general, weather systems move along an axis joining high and low isallobaric centers. Isallobaric gradients are important components of the wind as they increase or decrease the geostrophic wind.

An isopycnal is a line of constant density. An isoheight or isohypse is a line of constant geopotential height on a constant pressure surface chart.

Temperature and Related Subjects

An isotherm is a line that connects points on a map that have the same temperature. Therefore, all points through which an isotherm passes have the same or equal temperatures at the time indicated. An isotherm at 0 °C is called the freezing level. The term was coined by the Prussian geographer and natu-

ralist Alexander von Humboldt, who as part of his research into the geographical distribution of plants published the first map of isotherms in Paris, in 1817.

The 10 °C (50 °F) mean isotherm in July, marked by the red line, is commonly
used to define the border of the Arctic region

An isogeotherm is a line of equal mean annual temperature. An isocheim is a line of
equal mean winter temperature, and an isothere is a line of equal mean summer temperature.

An isohel is a line of equal or constant solar radi-ation.

Rainfall and Air Moisture

An isohyet or isohyetal line is a line joining points of equal rainfall on a map in a
given period . A map with isohyets is called an isohyetal map.

An isohume is a line of constant relative humidity, while a isodrosotherm is a line of
equal or constant dew point.

An isoneph is a line indicating equal cloud cover.

An isochalaz is a line of constant frequency of hail storms, and an isobront is a line
drawn through geographical points at which a given phase of thunderstorm activity
occurred simultaneously.

Snow cover is frequently shown as a contour-line map.

Wind

An isotach is a line joining points with constant wind speed. In meteorology, the term isogon refers to a line of constant wind direction.

Freeze and Thaw

An isopectic line denotes equal dates of ice formation each winter, and an isotac denotes equal dates of thawing.

Physical Geography and Oceanography

Elevation and Depth

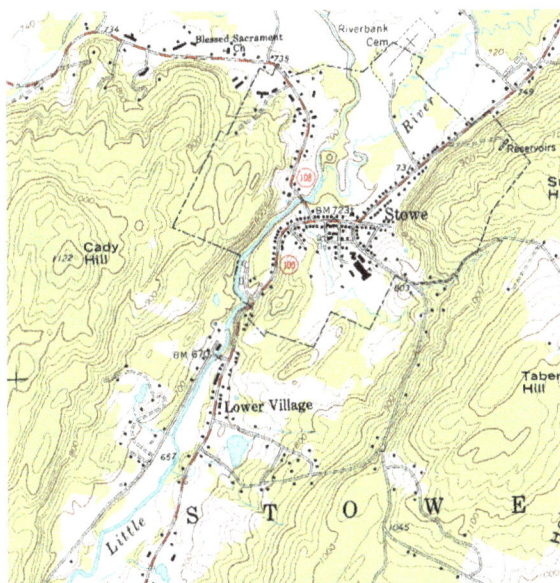

Topographic map of Stowe, Vermont. The brown contour lines represent the elevation. The contour interval is 20 feet.

Contours are one of several common methods used to denote elevation or altitude and depth on maps. From these contours, a sense of the general terrain can be determined. They are used at a variety of scales, from large-scale engineering drawings and architectural plans, through topographic maps and bathymetric charts, up to continental-scale maps.

"Contour line" is the most common usage in cartography, but isobath for underwater depths on bathymetric maps and isohypse for elevations are also used.

In cartography, the contour interval is the elevation difference between adjacent contour lines. The contour interval should be the same over a single map. When calculated as a ratio against the map scale, a sense of the hilliness of the terrain can be derived.

Interpretation

There are several rules to note when interpreting terrain contour lines:

- The rule of V's: sharp-pointed vees usually are in stream valleys, with the drainage channel passing through the point of the vee, with the vee pointing upstream. This is a consequence of erosion.

- The rule of O's: closed loops are normally uphill on the inside and downhill on the outside, and the innermost loop is the highest area. If a loop instead represents a depression, some maps note this by short lines radiating from the inside of the loop, called "hachures".

- Spacing of contours: close contours indicate a steep slope; distant contours a shallow slope. Two or more contour lines merging indicates a cliff. By counting the number of contours that cross a segment of a stream, you can approximate the stream gradient.

Of course, to determine differences in elevation between two points, the contour interval, or distance in altitude between two adjacent contour lines, must be known, and this is given at the bottom of the map. Usually contour intervals are consistent throughout a map, but there are exceptions. Sometimes intermediate contours are present in flatter areas; these can be dashed or dotted lines at half the noted contour interval. When contours are used with hypsometric tints on a small-scale map that includes mountains and flatter low-lying areas, it is common to have smaller intervals at lower elevations so that detail is shown in all areas. Conversely, for an island which consists of a plateau surrounded by steep cliffs, it is possible to use smaller intervals as the height increases.

Electrostatics

An isopotential map is a measure of electrostatic potential in space, often depicted in two dimensions with the electostatic charges inducing that electric potential. The term equipotential line or isopotential line refers to a curve of constant electric potential. Whether crossing an equipotential line represents ascending or descending the potential is inferred from the labels on the charges. In three dimensions, equipotential surfaces may be depicted with a two dimensional cross-section, showing equipotential lines at the intersection of the surfaces and the cross-section.

The general mathematical term level set is often used to describe the full collection of points having a particular potential, especially in higher dimensional space.

Magnetism

In the study of the Earth's magnetic field, the term isogon or isogonic line refers to a line of constant magnetic declination, the variation of magnetic north from geographic north. An agonic line is drawn through points of zero magnetic declination. An isoporic line refers to a line of constant annual variation of magnetic declination .

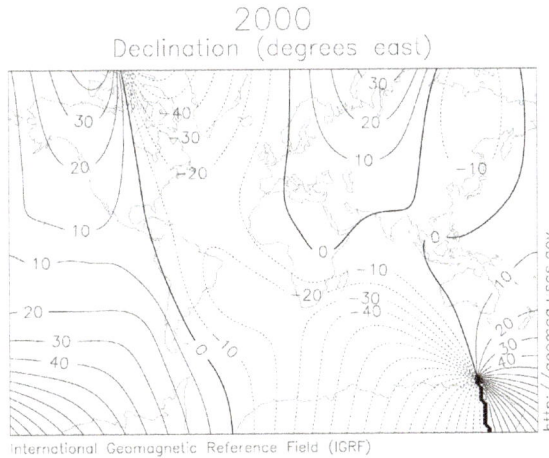

Isogonic lines for the year 2000. The agonic lines are thicker and labeled with "0".

An isoclinic line connects points of equal magnetic dip, and an aclinic line is the isoclinic line of magnetic dip zero.

An isodynamic line connects points with the same intensity of magnetic force.

Oceanography

Besides ocean depth, oceanographers use contour to describe diffuse variable phenomena much as meteorologists do with atmospheric phenomena. In particular, isobathytherms are lines showing depths of water with equal temperature, isohalines show lines of equal ocean salinity, and Isopycnals are surfaces of equal water density.

Geology

Various geological data are rendered as contour maps in structural geology, sedimentology, stratigraphy and economic geology. Contour maps are used to show the below ground surface of geologic strata, fault surfaces (especially low angle thrust faults) and unconformities. Isopach maps use isopachs (lines of equal thickness) to illustrate variations in thickness of geologic units.

Environmental Science

In discussing pollution, density maps can be very useful in indicating sources and areas of greatest contamination. Contour maps are especially useful for diffuse forms or scales of pollution. Acid precipitation is indicated on maps with isoplats. Some of the most widespread applications of environmental science contour maps involve mapping of environmental noise (where lines of equal sound pressure level are denoted isobels), air pollution, soil contamination, thermal pollution and groundwater contamination.

By contour planting and contour ploughing, the rate of water runoff and thus soil erosion can be substantially reduced; this is especially important in riparian zones.

Ecology

An isoflor is an isopleth contour connecting areas of comparable biological diversity. Usually, the variable is the number of species of a given genus or family that occurs in a region. Isoflor maps are thus used to show distribution patterns and trends such as centres of diversity.

Social Sciences

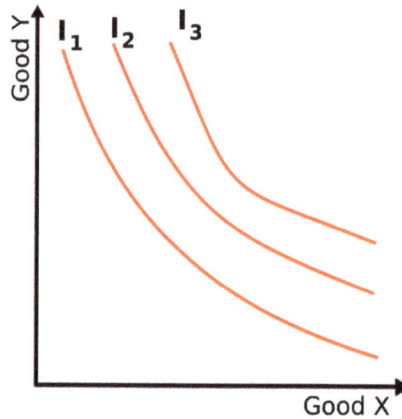

From economics, an indifference map with three indifference curves shown. All points on a particular indifference curve have the same value of the utility function, whose values implicitly come out of the page in the unshown third dimension.

In economics, contour lines can be used to describe features which vary quantitatively over space. An isochrone shows lines of equivalent drive time or travel time to a given location and is used in the generation of isochrone maps. An isotim shows equivalent transport costs from the source of a raw material, and an isodapane shows equivalent cost of travel time.

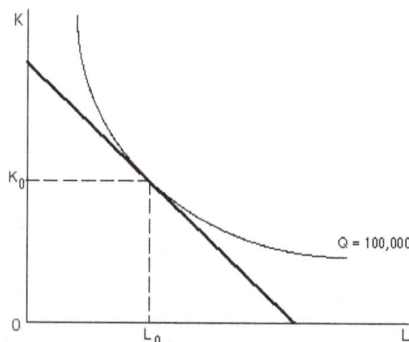

A single production isoquant (convex) and a single isocost curve (linear). Labor usage is plotted horizontally and physical capital usage is plotted vertically.

Contour lines are also used to display non-geographic information in economics. In-difference curves (as shown at left) are used to show bundles of goods to which a person would assign equal utility. An isoquant (in the image at right) is a curve of equal production quantity for alternative combinations of input usages, and an isocost curve (also in the image at right) shows alternative combinations of input usages having equal production costs.

In political science an analogous method is used in understanding coalitions (for example the diagram in Laver and Shepsle's work).

In population dynamics, an isocline shows the set of population sizes at which the rate of change, or partial derivative, for one population in a pair of interacting populations is zero.

Statistics

In statistics, isodensity lines or isodensanes are lines that joint points with the same probability density. Isodensanes are used to display bivariate distributions.

Thermodynamics, Engineering, and Other Sciences

Various types of graphs in thermodynamics, engineering, and other sciences use isobars (constant pressure), isotherms (constant temperature), isochors (constant specific volume), or other types of isolines, even though these graphs are usually not related to maps. Such isolines are useful for representing more than two dimensions (or quantities) on two-dimensional graphs. Common examples in thermodynamics are some types of phase diagrams.

Isoclines are used to solve ordinary differential equations.

In interpreting radar images, an isodop is a line of equal Doppler velocity, and an isoecho is a line of equal radar reflectivity.

Other Phenomena

- isochasm: aurora equal occurrence
- isochor: volume
- isodose: Absorbed dose of radiation
- isophene: biological events occurring with coincidence such as plants flowering
- isophote: illuminance

History

The idea of lines that join points of equal value was rediscovered several times. In 1701,

Edmond Halley used such lines (isogons) on a chart of magnetic variation. The Dutch engineer Nicholas Cruquius drew the bed of the river Merwede with lines of equal depth (isobaths) at intervals of 1 fathom in 1727, and Philippe Buache used them at 10-fathom intervals on a chart of the English Channel that was prepared in 1737 and published in 1752. Such lines were used to describe a land surface (contour lines) in a map of the Duchy of Modena and Reggio by Domenico Vandelli in 1746, and they were studied theoretically by Ducarla in 1771, and Charles Hutton used them in the Schiehallion experiment. In 1791, a map of France by J. L. Dupain-Triel used contour lines at 20-metre intervals, hachures, spot-heights and a vertical section. In 1801, the chief of the Corps of Engineers, Haxo, used contour lines at the larger scale of 1:500 on a plan of his projects for Rocca d'Aufo.

By around 1843, when the Ordnance Survey started to regularly record contour lines in Great Britain and Ireland, they were already in general use in European countries. Isobaths were not routinely used on nautical charts until those of Russia from 1834, and those of Britain from 1838.

When maps with contour lines became common, the idea spread to other applications. Perhaps the latest to develop are air quality and noise pollution contour maps, which first appeared in the US, in approximately 1970, largely as a result of national legislation requiring spatial delineation of these parameters. In 2007, Pictometry International was the first to allow users to dynamically generate elevation contour lines to be laid over oblique images.

Graphical Design

To maximize readability of contour maps, there are several design choices available to the map creator, principally line weight, line color, line type and method of numerical marking.

Line weight is simply the darkness or thickness of the line used. This choice is made based upon the least intrusive form of contours that enable the reader to decipher the background information in the map itself. If there is little or no content on the base map, the contour lines may be drawn with relatively heavy thickness. Also, for many forms of contours such as topographic maps, it is common to vary the line weight and/or color, so that a different line characteristic occurs for certain numerical values. For example, in the topographic map above, the even hundred foot elevations are shown in a different weight from the twenty foot intervals.

Line color is the choice of any number of pigments that suit the display. Sometimes a sheen or gloss is used as well as color to set the contour lines apart from the base map. Line colour can be varied to show other information.

Line type refers to whether the basic contour line is solid, dashed, dotted or broken in some other pattern to create the desired effect. Dotted or dashed lines are often used when the underlying base map conveys very important (or difficult to read) information. Broken line types are used when the location of the contour line is inferred.

Numerical marking is the manner of denoting the arithmetical values of contour lines. This can be done by placing numbers along some of the contour lines, typically using interpolation for intervening lines. Alternatively a map key can be produced associating the contours with their values.

If the contour lines are not numerically labeled and adjacent lines have the same style (with the same weight, color and type), then the direction of the gradient cannot be determined from the contour lines alone. However, if the contour lines cycle through three or more styles, then the direction of the gradient can be determined from the lines. The orientation of the numerical text labels is often used to indicate the direction of the slope.

Plan View Versus Profile View

Most commonly contour lines are drawn in plan view, or as an observer in space would view the Earth's surface: ordinary map form. However, some parameters can often be displayed in profile view showing a vertical profile of the parameter mapped. Some of the most common parameters mapped in profile are air pollutant concentrations and sound levels. In each of those cases it may be important to analyze (air pollutant concentrations or sound levels) at varying heights so as to determine the air quality or noise health effects on people at different elevations, for example, living on different floor levels of an urban apartment. In actuality, both plan and profile view contour maps are used in air pollution and noise pollution studies.

Labeling Contour Maps

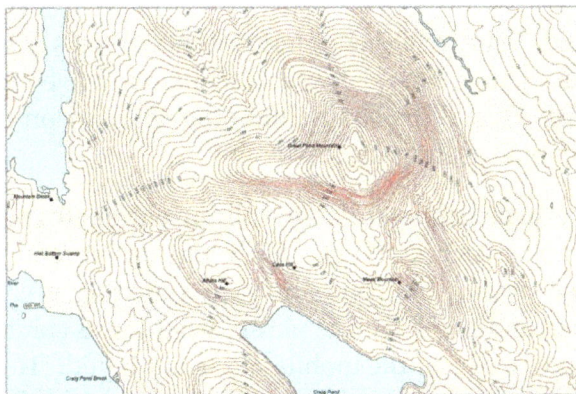

Contour map labeled aesthetically in an "elevation up" manner.

Labels are a critical component of elevation maps. A properly labeled contour map helps the reader to quickly interpret the shape of the terrain. If numbers are placed close to each other, it means that the terrain is steep. Labels should be placed along a slightly curved line "pointing" to the summit or nadir, from several directions if possible, making the visual identification of the summit or nadir easy. Contour labels can be oriented so a reader is facing uphill when reading the label.

Manual labeling of contour maps is a time-consuming process, however, there are a few software systems that can do the job automatically and in accordance with cartographic conventions, called automatic label placement.

Gas Chromatography

Gas chromatography (GC) is a common type of chromatography used in analytical chemistry for separating and analyzing compounds that can be vaporized without decomposition. Typical uses of GC include testing the purity of a particular substance, or separating the different components of a mixture (the relative amounts of such components can also be determined). In some situations, GC may help in identifying a compound. In preparative chromatography, GC can be used to prepare pure compounds from a mixture.

In gas chromatography, the *mobile phase* (or "moving phase") is a carrier gas, usually an inert gas such as helium or an unreactive gas such as nitrogen. Helium remains the most commonly used carrier gas in about 90% of instruments although hydrogen is preferred for improved separations. The *stationary phase* is a microscopic layer of liquid or polymer on an inert solid support, inside a piece of glass or metal tubing called a column (an homage to the fractionating column used in distillation). The instrument used to perform gas chromatography is called a *gas chromatograph* (or "aerograph", "gas separator").

The gaseous compounds being analyzed interact with the walls of the column, which is coated with a stationary phase. This causes each compound to elute at a different time, known as the *retention time* of the compound. The comparison of retention times is what gives GC its analytical usefulness.

Gas chromatography is in principle similar to column chromatography (as well as other forms of chromatography, such as HPLC, TLC), but has several notable differences. First, the process of separating the compounds in a mixture is carried out between a liquid stationary phase and a gas mobile phase, whereas in column chromatography the stationary phase is a solid and the mobile phase is a liquid. (Hence the full name of the procedure is "Gas–liquid chromatography", referring to the mobile and stationary phases, respectively.) Second, the column through which the gas phase passes is located in an oven where the temperature of the gas can be controlled, whereas column chromatography (typically) has no such temperature control. Finally, the concentration of a compound in the gas phase is solely a function of the vapor pressure of the gas.

Gas chromatography is also similar to fractional distillation, since both processes separate the components of a mixture primarily based on boiling point (or vapor pressure) differences. However, fractional distillation is typically used to separate components of a mixture on a large scale, whereas GC can be used on a much smaller scale (i.e. microscale).

Gas chromatography is also sometimes known as vapor-phase chromatography (VPC), or gas–liquid partition chromatography (GLPC). These alternative names, as well as their respective abbreviations, are frequently used in scientific literature. Strictly speaking, GLPC is the most correct terminology, and is thus preferred by many authors.

History

Chromatography dates to 1903 in the work of the Russian scientist, Mikhail Semenovich Tswett. German graduate student Fritz Prior developed solid state gas chromatography in 1947. Archer John Porter Martin, who was awarded the Nobel Prize for his work in developing liquid–liquid (1941) and paper (1944) chromatography, laid the foundation for the development of gas chromatography and he later produced liquid-gas chromatography (1950). Erika Cremer laid the groundwork, and oversaw much of Prior's work.

GC Analysis

A gas chromatograph is a chemical analysis instrument for separating chemicals in a complex sample. A gas chromatograph uses a flow-through narrow tube known as the *column*, through which different chemical constituents of a sample pass in a gas stream (carrier gas, *mobile phase*) at different rates depending on their various chemical and physical properties and their interaction with a specific column filling, called the *stationary phase*. As the chemicals exit the end of the column, they are detected and identified electronically. The function of the stationary phase in the column is to separate different components, causing each one to exit the column at a different time (*retention time*). Other parameters that can be used to alter the order or time of retention are the carrier gas flow rate, column length and the temperature.

In a GC analysis, a known volume of gaseous or liquid analyte is injected into the "entrance" (head) of the column, usually using a microsyringe (or, solid phase microextraction fibers, or a gas source switching system). As the carrier gas sweeps the analyte molecules through the column, this motion is inhibited by the adsorption of the analyte molecules either onto the column walls or onto packing materials in the column. The rate at which the molecules progress along the column depends on the strength of adsorption, which in turn depends on the type of molecule and on the stationary phase materials. Since each type of molecule has a different rate of progression, the various components of the analyte mixture are separated as they progress along the column and reach the end of the column at different times (retention time). A detector is used to monitor the outlet stream from the column; thus, the time at which each component reaches the outlet and the amount of that component can be determined. Generally, substances are identified (qualitatively) by the order in which they emerge (elute) from the column and by the retention time of the analyte in the column.

Physical Components

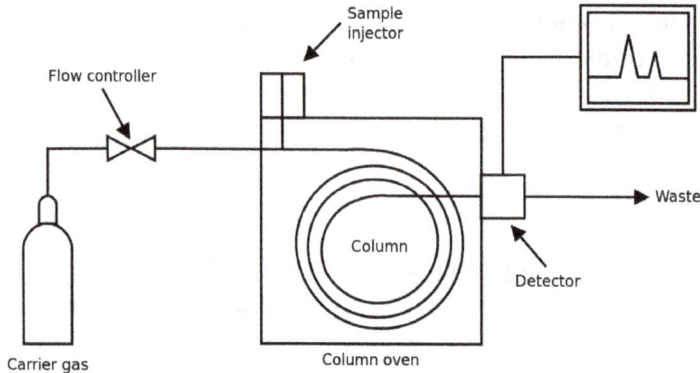

Diagram of a gas chromatograph.

Autosamplers

The autosampler provides the means to introduce a sample automatically into the inlets. Manual insertion of the sample is possible but is no longer common. Automatic insertion provides better reproducibility and time-optimization.

Different kinds of autosamplers exist. Autosamplers can be classified in relation to sample capacity (auto-injectors vs. autosamplers, where auto-injectors can work a small number of samples), to robotic technologies (XYZ robot vs. rotating robot – the most common), or to analysis:

- Liquid
- Static head-space by syringe technology
- Dynamic head-space by transfer-line technology
- Solid phase microextraction (SPME)

Traditionally autosampler manufacturers are different from GC manufacturers and currently no GC manufacturer offers a complete range of autosamplers. Historically, the countries most active in autosampler technology development are the United States, Italy, Switzerland, and the United Kingdom.

Inlets

The column inlet (or injector) provides the means to introduce a sample into a continuous flow of carrier gas. The inlet is a piece of hardware attached to the column head.

Common inlet types are:

- S/SL (split/splitless) injector; a sample is introduced into a heated small chamber via a syringe through a septum – the heat facilitates volatilization of the sample and sample matrix. The carrier gas then either sweeps the entirety (splitless mode) or a portion (split mode) of the sample into the column. In split mode, a part of the sample/carrier gas mixture in the injection chamber is exhausted through the split vent. Split injection is preferred when working with samples with high analyte concentrations (>0.1%) whereas splitless injection is best suited for trace analysis with low amounts of analytes (<0.01%). In split-less mode the split valve opens after a pre-set amount of time to purge heavier elements that would otherwise contaminate the system. This pre-set (splitless) time should be optimized, the shorter time (e.g., 0.2 min) ensures less tailing but loss in response, the longer time (2 min) increases tailing but also signal.

- On-column inlet; the sample is here introduced directly into the column in its entirety without heat, or at a temperature below the boiling point of the solvent. The low temperature condenses the sample into a narrow zone. The column and inlet can then be heated, releasing the sample into the gas phase. This ensures the lowest possible temperature for chromatography and keeps samples from decomposing above their boiling point.

- PTV injector; Temperature-programmed sample introduction was first described by Vogt in 1979. Originally Vogt developed the technique as a method for the introduction of large sample volumes (up to 250 µL) in capillary GC. Vogt introduced the sample into the liner at a controlled injection rate. The temperature of the liner was chosen slightly below the boiling point of the solvent. The low-boiling solvent was continuously evaporated and vented through the split line. Based on this technique, Poy developed the programmed temperature vaporising injector; PTV. By introducing the sample at a low initial liner temperature many of the disadvantages of the classic hot injection techniques could be circumvented.

- Gas source inlet or gas switching valve; gaseous samples in collection bottles are connected to what is most commonly a six-port switching valve. The carrier gas flow is not interrupted while a sample can be expanded into a previously evacuated sample loop. Upon switching, the contents of the sample loop are inserted into the carrier gas stream.

- P/T (Purge-and-Trap) system; An inert gas is bubbled through an aqueous sample causing insoluble volatile chemicals to be purged from the matrix. The volatiles are 'trapped' on an absorbent column (known as a trap or concentrator) at ambient temperature. The trap is then heated and the volatiles are directed into the carrier gas stream. Samples requiring preconcentration or purification can be introduced via such a system, usually hooked up to the S/SL port.

The choice of carrier gas (mobile phase) is important. Hydrogen has a range of flow rates that are comparable to helium in efficiency. However, helium may be more efficient and provide the best separation if flow rates are optimized. Helium is non-flammable and works with a greater number of detectors and older instruments. Therefore, helium is the most common carrier gas used. However, the price of helium has gone up considerably over recent years, causing an increasing number of chromatographers to switch to hydrogen gas. Historical use, rather than rational consideration, may contribute to the continued preferential use of helium.

Detectors

The most commonly used detectors are the flame ionization detector (FID) and the thermal conductivity detector (TCD). Both are sensitive to a wide range of components, and both work over a wide range of concentrations. While TCDs are essentially universal and can be used to detect any component other than the carrier gas (as long as their thermal conductivities are different from that of the carrier gas, at detector temperature), FIDs are sensitive primarily to hydrocarbons, and are more sensitive to them than TCD. However, a FID cannot detect water. Both detectors are also quite robust. Since TCD is non-destructive, it can be operated in-series before a FID (destructive), thus providing complementary detection of the same analytes.

Other detectors are sensitive only to specific types of substances, or work well only in narrower ranges of concentrations. They include:

- Thermal Conductivity detector (TCD), this common detector relies on the thermal conductivity of matter passing around a tungsten -rhenium filament with a current traveling through it. In this set up helium or nitrogen serve as the carrier gas because of their relatively high thermal conductivity which keep the filament cool and maintain uniform resistivity and electrical efficiency of the filament. However, when analyte molecules elute from the column, mixed with carrier gas, the thermal conductivity decreases and this causes a detector response. The response is due to the decreased thermal conductivity causing an increase in filament temperature and resistivity resulting in fluctuations in voltage. Detector sensitivity is proportional to filament current while it is inversely proportional to the immediate environmental temperature of that detector as well as flow rate of the carrier gas.

- Flame Ionization detector (FID), in this common detector electrodes are placed adjacent to a flame fueled by hydrogen / air near the exit of the column, and when carbon containing compounds exit the column they are pyrolyzed by the flame. This detector works only for organic / hydrocarbon containing compounds due to the ability of the carbons to form cations and electrons upon pyrolysis which generates a current between the electrodes. The increase in current is translated and appears as a peak in a chromatogram. FIDs have low detection limits (a few picograms per second) but they are unable to generate ions from carbonyl containing carbons. FID compatible carrier gasses include nitrogen, helium, and argon.

- Catalytic combustion detector (CCD), which measures combustible hydrocarbons and hydrogen.

- Discharge ionization detector (DID), which uses a high-voltage electric discharge to produce ions.

- Dry electrolytic conductivity detector (DELCD), which uses an air phase and high temperature (v. Coulsen) to measure chlorinated compounds.

- Electron capture detector (ECD), which uses a radioactive beta particle (electron) source to measure the degree of electron capture. ECD are used for the detection of molecules containing electronegative / withdrawing elements and functional groups like halogens, carbonyl, nitriles, nitro groups, and organometalics. In this type of detector either nitrogen or 5% methane in argon is used as the mobile phase carrier gas. The carrier gas passes between two electrodes placed at the end of the column, and adjacent to the anode (negative electrode) resides a radioactive foil such as 63Ni. The radioactive foil emits a beta particle (electron) which collides with and ionizes the carrier gas to generate more ions resulting in a current. When analyte molecules with electronegative / withdrawing elements or functional groups electrons are captured which results in a decrease in current generating a detector response.

- Flame photometric detector (FPD), which uses a photomultiplier tube to detect spectral lines of the compounds as they are burned in a flame. Compounds eluting off the column are carried into a hydrogen fueled flame which excites specific elements in the molecules, and the excited elements (P,S, Halogens, Some Metals) emit light of specific characteristic wavelengths. The emitted light is filtered and detected by a photomultiplier tube. In particular, phosphorus emission is around 510–536 nm and sulfur emission os at 394 nm.

- Atomic Emission Detector (AED), a sample eluting from a column enters a chamber which is energized by microwaves that induce a plasma. The plasma causes the analyte sample to decompose and certain elements generate an atomic emission spectra. The atomic emission spectra is diffracted by a diffraction grating and detected by a series of photomultiplier tubes or photo diodes.

- Hall electrolytic conductivity detector (ElCD)

- Helium ionization detector (HID)

- Nitrogen–phosphorus detector (NPD), a form of thermionic detector where nitrogen and phosphorus alter the work function on a specially coated bead and a resulting current is measured.

- Alkali Flame Detector, AFD or Alkali Flame Ionization Detector, AFID. AFD has high sensitivity to nitrogen and phosphorus, similar to NPD. However, the alkaline metal ions are supplied with the hydrogen gas, rather than a bead above the flame. For this reason AFD does not suffer the "fatigue" of he NPD, but provides a constant sensitivity over long period of time. In addition, when alkali ions are not added to the flame, AFD operates like a standard FID.

- Infrared detector (IRD)

- Mass spectrometer (MS), also called GC-MS; highly effective and sensitive, even in a small quantity of sample.

- Photo-ionization detector (PID)

- The Polyarc reactor is an add-on to new or existing GC-FID instruments that converts all organic compounds to methane molecules prior to their detection by the FID. This technique can be used to improve the response of the FID and allow for the detection of many more carbon-containing compounds. The complete conversion of compounds to methane and the now equivalent response in the detector also eliminates the need for calibrations and standards because response factors are all equivalent to those of methane. This allows for the rapid analysis of complex mixtures that contain molecules where standards are not available.

- Pulsed discharge ionization detector (PDD)

- Thermionic ionization detector (TID)

- Vacuum Ultraviolet (VUV) represents the most recent development in Gas Chromatography detectors. Most chemical species absorb and have unique gas phase absorption cross sections in the approximately 120–240 nm VUV wavelength range monitored. Where absorption cross sections are known for analytes, the VUV detector is capable of absolute determination (without calibration) of the number of molecules present in the flow cell in the absence of chemical interferences.

Some gas chromatographs are connected to a mass spectrometer which acts as the detector. The combination is known as GC-MS. Some GC-MS are connected to an NMR spectrometer which acts as a backup detector. This combination is known as GC-MS-NMR. Some GC-MS-NMR are connected to an infrared spectrophotometer which acts as a backup detector. This combination is known as GC-MS-NMR-IR. It must, however, be stressed this is very rare as most analyses needed can be concluded via purely GC-MS.

Methods

The method is the collection of conditions in which the GC operates for a given analysis. Method development is the process of determining what conditions are adequate and/or ideal for the analysis required.

This image above shows the interior of a GeoStrata Technologies Eclipse Gas Chromatograph that runs continuously in three-minute cycles. Two valves are used to switch the test gas into the sample loop. After filling the sample loop with test gas, the valves are switched again applying carrier gas pressure to the sample loop and forcing the sample through the column for separation.

Conditions which can be varied to accommodate a required analysis include inlet tem-perature, detector temperature, column temperature and temperature program, carrier gas and carrier gas flow rates, the column's stationary phase, diameter and length, inlet type and flow rates, sample size and injection technique. Depending on the detector(s) installed on the GC, there may be a number of detector conditions that can also be varied. Some GCs also include valves which can change the route of sample and carrier flow. The timing of the opening and closing of these valves can be important to method development.

Carrier Gas Selection and Flow Rates

Typical carrier gases include helium, nitrogen, argon, hydrogen and air. Which gas to use is usually determined by the detector being used, for example, a DID requires helium as the carrier gas. When analyzing gas samples, however, the carrier is sometimes selected based on the sample's matrix, for example, when analyzing a mixture in argon, an argon carrier is preferred, because the argon in the sample does not show up on the chromatogram. Safety and availability can also influence carrier selection, for example, hydrogen is flammable, and high-purity helium can be difficult to obtain in some areas of the world. As a result of helium becoming more scarce, hydrogen is often being substituted for helium as a carrier gas in several applications.

The purity of the carrier gas is also frequently determined by the detector, though the level of sensitivity needed can also play a significant role. Typically, purities of 99.995% or higher are used. The most common purity grades required by modern instruments for the majority of sensitivities are 5.0 grades, or 99.999% pure meaning that there is a total of 10ppm of impurities in the carrier gas that could affect the results. The highest purity grades in common use are 6.0 grades, but the need for detection at very low levels in some forensic and environmental applications has driven the need for carrier gases at 7.0 grade purity and these are now commercially available. Trade names for typical purities include "Zero Grade," "Ultra-High Purity (UHP) Grade," "4.5 Grade" and "5.0 Grade."

The carrier gas linear velocity affects the analysis in the same way that temperature does. The higher the linear velocity the faster the analysis, but the lower the separation between analytes. Selecting the linear velocity is therefore the same com-promise between the level of separation and length of analysis as selecting the column temperature. The linear velocity will be implemented by means of the carrier gas flow rate, with regards to the inner diameter of the column.

With GCs made before the 1990s, carrier flow rate was controlled indirectly by controlling the carrier inlet pressure, or "column head pressure." The actual flow rate was measured at the outlet of the column or the detector with an electronic flow meter, or a bubble flow meter, and could be an involved, time consuming, and frustrating process. The pressure setting was not able to be varied during the run, and thus the flow was essentially constant during the analysis. The relation between flow rate and inlet pressure is calculated with Poiseuille's equation for compressible fluids.

Many modern GCs, however, electronically measure the flow rate, and electronically control the carrier gas pressure to set the flow rate. Consequently, carrier pressures and flow rates can be adjusted during the run, creating pressure/flow programs similar to temperature programs.

Stationary Compound Selection

The polarity of the solute is crucial for the choice of stationary compound, which in an optimal case would have a similar polarity as the solute. Common stationary phases in open tubular columns are cyanopropylphenyl dimethyl polysiloxane, carbowax polyethyleneglycol, biscyanopropyl cyanopropylphenyl polysiloxane and diphenyl dimethyl polysiloxane. For packed columns more options are available.

Inlet Types and Flow Rates

The choice of inlet type and injection technique depends on if the sample is in liquid, gas, adsorbed, or solid form, and on whether a solvent matrix is present that has to be vaporized. Dissolved samples can be introduced directly onto the column via a COC injector, if the conditions are well known; if a solvent matrix has to be vaporized and partially removed, a S/SL injector is used (most common injection technique); gaseous samples (e.g., air cylinders) are usually injected using a gas switching valve system; adsorbed samples (e.g., on adsorbent tubes) are introduced using either an external (online or off-line) desorption apparatus such as a purge-and-trap system, or are desorbed in the injector (SPME applications).

Sample Size and Injection Technique

Sample Injection

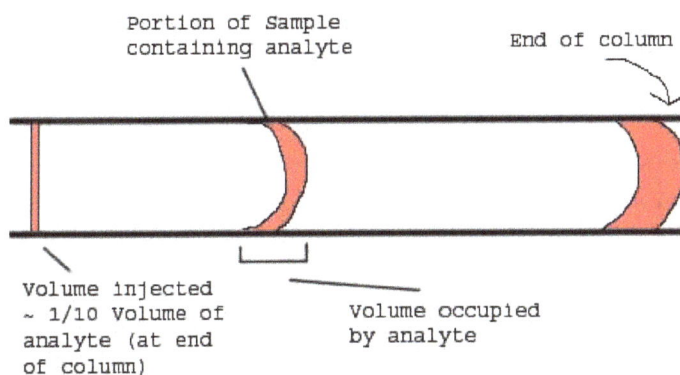

The rule of ten in gas chromatography

The real chromatographic analysis starts with the introduction of the sample onto the column. The development of capillary gas chromatography resulted in many practical problems with the injection technique. The technique of on-column injection, often used with packed columns, is usually not possible with capillary columns. The injection system in the capillary gas chromatograph should fulfil the following two requirements:

1. The amount injected should not overload the column.

2. The width of the injected plug should be small compared to the spreading due to the chromatographic process. Failure to comply with this requirement will reduce the separation capability of the column. As a general rule, the volume injected, V_{inj}, and the volume of the detector cell, V_{det}, should be about 1/10 of the volume occupied by the portion of sample containing the molecules of interest (analytes) when they exit the column.

Some general requirements which a good injection technique should fulfill are:

- It should be possible to obtain the column's optimum separation efficiency.

- It should allow accurate and reproducible injections of small amounts of representative samples.

- It should induce no change in sample composition. It should not exhibit discrimination based on differences in boiling point, polarity, concentration or thermal/catalytic stability.

- It should be applicable for trace analysis as well as for undiluted samples.

However, there are a number of problems inherent in the use of syringes for injection, even when they are not damaged:

- Even the best syringes claim an accuracy of only 3%, and in unskilled hands, errors are much larger

- The needle may cut small pieces of rubber from the septum as it injects sample through it. These can block the needle and prevent the syringe filling the next time it is used. It may not be obvious of what happened.

- A fraction of the sample may get trapped in the rubber, to be released during subsequent injections. This can give rise to ghost peaks in the chromatogram.

- There may be selective loss of the more volatile components of the sample by evaporation from the tip of the needle.

Column Selection

The choice of column depends on the sample and the active measured. The main chemical attribute regarded when choosing a column is the polarity of the mixture, but functional groups can play a large part in column selection. The polarity of the sample must closely match the polarity of the column stationary phase to increase resolution and separation while reducing run time. The separation and run time also depends on the film thickness (of the stationary phase), the column diameter and the column length.

Column Temperature and Temperature Program

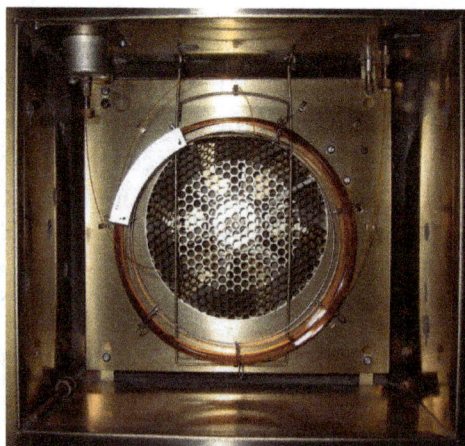

A gas chromatography oven, open to show a capillary column

The column(s) in a GC are contained in an oven, the temperature of which is precisely controlled electronically. (When discussing the "temperature of the column," an analyst is technically referring to the temperature of the column oven. The distinction, however, is not important and will not subsequently be made in this article.)

The rate at which a sample passes through the column is directly proportional to the temperature of the column. The higher the column temperature, the faster the sample moves through the column. However, the faster a sample moves through the column, the less it interacts with the stationary phase, and the less the analytes are separated.

In general, the column temperature is selected to compromise between the length of the analysis and the level of separation.

A method which holds the column at the same temperature for the entire analysis is called "isothermal." Most methods, however, increase the column temperature during the analysis, the initial temperature, rate of temperature increase (the temperature "ramp"), and final temperature are called the "temperature program."

A temperature program allows analytes that elute early in the analysis to separate adequately, while shortening the time it takes for late-eluting analytes to pass through the column.

Data Reduction and Analysis

Qualitative Analysis

Generally chromatographic data is presented as a graph of detector response (y-axis) against retention time (x-axis), which is called a chromatogram. This provides a spectrum of peaks for a sample representing the analytes present in a sample eluting from the column at different times. Retention time can be used to identify analytes if the method conditions are constant. Also, the pattern of peaks will be constant for a sample under constant conditions and can identify complex mixtures of analytes. However, in most modern applications, the GC is connected to a mass spectrometer or similar detector that is capable of identifying the analytes represented by the peaks.

Quantitative Analysis

The area under a peak is proportional to the amount of analyte present in the chromatogram. By calculating the area of the peak using the mathematical function of integration, the concentration of an analyte in the original sample can be determined. Concentration can be calculated using a calibration curve created by finding the response for a series of concentrations of analyte, or by determining the relative response factor of an analyte. The relative response factor is the expected ratio of an analyte to an internal standard (or external standard) and is calculated by finding the response of a known amount of analyte and a constant amount of internal standard (a chemical added to the sample at a constant concentration, with a distinct retention time to the analyte).

In most modern GC-MS systems, computer software is used to draw and integrate peaks, and match MS spectra to library spectra.

Applications

In general, substances that vaporize below 300 °C (and therefore are stable up to that temperature) can be measured quantitatively. The samples are also required to be salt-free; they should not contain ions. Very minute amounts of a substance can be measured, but it is often required that the sample must be measured in comparison to a sample containing the pure, suspected substance known as a reference standard.

Various temperature programs can be used to make the readings more meaningful; for example to differentiate between substances that behave similarly during the GC process.

Professionals working with GC analyze the content of a chemical product, for example in assuring the quality of products in the chemical industry; or measuring toxic substances in soil, air or water. GC is very accurate if used properly and can measure picomoles of a substance in a 1 ml liquid sample, or parts-per-billion concentrations in gaseous samples.

In practical courses at colleges, students sometimes get acquainted to the GC by studying the contents of Lavender oil or measuring the ethylene that is secreted by *Nicotiana benthamiana* plants after artificially injuring their leaves. These GC analyse hydrocarbons (C2-C40+). In a typical experiment, a packed column is used to separate the light gases, which are then detected with a TCD. The hydrocarbons are separated using a capillary column and detected with a FID. A complication with light gas analyses that include H_2 is that He, which is the most common and most sensitive inert carrier (sensitivity is proportional to molecular mass) has an almost identical thermal conductivity to hydrogen (it is the difference in thermal conductivity between two separate filaments in a Wheatstone Bridge type arrangement that shows when a component has been eluted). For this reason, dual TCD instruments used with a separate channel for hydrogen that uses nitrogen as a carrier are common. Argon is often used when analysing gas phase chemistry reactions such as F-T synthesis so that a single carrier gas can be used rather than two separate ones. The sensitivity is less, but this is a trade off for simplicity in the gas supply.

Gas Chromatography is used extensively in forensic science. Disciplines as diverse as solid drug dose (pre-consumption form) identification and quantification, arson investigation, paint chip analysis, and toxicology cases, employ GC to identify and quantify various biological specimens and crime-scene evidence.

GCs in Popular Culture

Movies, books and TV shows tend to misrepresent the capabilities of gas chromatography and the work done with these instruments.

In the U.S. TV show CSI, for example, GCs are used to rapidly identify unknown samples. For example, an analyst may say fifteen minutes after receiving the sample: "This is gasoline bought at a Chevron station in the past two weeks."

In fact, a typical GC analysis takes much more time; sometimes a single sample must be run more than an hour according to the chosen program; and even more time is needed to "heat out" the column so it is free from the first sample and can be used for the next. Equally, several runs are needed to confirm the results of a study – a GC analysis of a single sample may simply yield a result per chance.

Also, GC does not positively identify most samples; and not all substances in a sample will necessarily be detected. All a GC truly tells you is at which relative time a component eluted from the column and that the detector was sensitive to it. To make results meaningful, analysts need to know which components at which concentrations are to be expected; and even then a small amount of a substance can hide itself behind a substance having both a higher concentration and the same relative elution time. Last but not least it is often needed to check the results of the sample against a GC analysis of a reference sample containing only the suspected substance.

A GC-MS can remove much of this ambiguity, since the mass spectrometer will identify the component's molecular weight. But this still takes time and skill to do properly.

Similarly, most GC analyses are not push-button operations. You cannot simply drop a sample vial into an auto-sampler's tray, push a button and have a computer tell you everything you need to know about the sample. The operating program must be carefully chosen according to the expected sample composition.

A push-button operation can exist for running similar samples repeatedly, such as in a chemical production environment or for comparing 20 samples from the same experiment to calculate the mean content of the same substance. However, for the kind of investigative work portrayed in books, movies and TV shows this is clearly not the case.

Mass Spectrometry

Mass spectrometry (MS) is an analytical technique that ionizes chemical species and sorts the ions based on their mass to charge ratio. In simpler terms, a mass spectrum measures the masses within a sample. Mass spectrometry is used in many different fields and is applied to pure samples as well as complex mixtures.

SIMS mass spectrometer, model IMS 3f.

Orbitrap mass spectrometer.

A mass spectrum is a plot of the ion signal as a function of the mass-to-charge ratio. These spectra are used to determine the elemental or isotopic signature of a sample, the masses of particles and of molecules, and to elucidate the chemical structures of molecules, such as peptides and other chemical compounds.

In a typical MS procedure, a sample, which may be solid, liquid, or gas, is ionized, for example by bombarding it with electrons. This may cause some of the sample's molecules to break into charged fragments. These ions are then separated according to their mass-to-charge ratio, typically by accelerating them and subjecting them to an electric or magnetic field: ions of the same mass-to-charge ratio will undergo the same amount of deflection. The ions are detected by a mechanism capable of detecting charged particles, such as an electron multiplier. Results are displayed as spectra of the relative abundance of detected ions as a function of the mass-to-charge ratio. The atoms or molecules in the sample can be identified by correlating known masses to the identified masses or through a characteristic fragmentation pattern.

History

In 1886, Eugen Goldstein observed rays in gas discharges under low pressure that traveled away from the anode and through channels in a perforated cathode, opposite to the direction of negatively charged cathode rays (which travel from cathode to anode).

Goldstein called these positively charged anode rays "Kanalstrahlen"; the standard translation of this term into English is "canal rays". Wilhelm Wien found that strong electric or magnetic fields deflected the canal rays and, in 1899, constructed a device with parallel electric and magnetic fields that separated the positive rays according to their charge-to-mass ratio (Q/m). Wien found that the charge-to-mass ratio depended on the nature of the gas in the discharge tube. English scientist J.J. Thomson later improved on the work of Wien by reducing the pressure to create the mass spectrograph.

Replica of J.J. Thompson's third mass spectrometer.

Calutron mass spectrometers were used in the Manhattan Project for uranium enrichment.

The word *spectrograph* had become part of the international scientific vocabulary by 1884. The linguistic roots are a combination and removal of bound morphemes and free morphemes which relate to the terms *spectr-um* and *phot-ograph-ic plate*. Early *spectrometry* devices that measured the mass-to-charge ratio of ions were called *mass spectrographs* which consisted of instruments that recorded a spectrum of mass values on a photographic plate. A *mass spectroscope* is similar to a *mass spectrograph* except that the beam of ions is directed onto a phosphor screen. A mass spectroscope configuration was used in early instruments when it was desired that the effects of adjustments be quickly observed. Once the instrument was properly adjusted, a photographic plate was inserted and exposed. The term mass spectroscope continued to be used even though the direct illumination of a phosphor screen was replaced by indirect measurements with an oscilloscope. The use of the term *mass spectroscopy* is now discouraged due to the possibility of confusion with light spectroscopy. Mass spectrometry is often abbreviated as *mass-spec* or simply as *MS*.

Modern techniques of mass spectrometry were devised by Arthur Jeffrey Dempster and F.W. Aston in 1918 and 1919 respectively.

Sector mass spectrometers known as calutrons were used for separating the isotopes of uranium developed by Ernest O. Lawrence during the Manhattan Project. Calutron mass spectrometers were used for uranium enrichment at the Oak Ridge, Tennessee Y-12 plant established during World War II.

In 1989, half of the Nobel Prize in Physics was awarded to Hans Dehmelt and Wolfgang Paul for the development of the ion trap technique in the 1950s and 1960s.

In 2002, the Nobel Prize in Chemistry was awarded to John Bennett Fenn for the development of electrospray ionization (ESI) and Koichi Tanaka for the development of soft laser desorption (SLD) and their application to the ionization of biological macromolecules, especially proteins.

Parts of a Mass Spectrometer

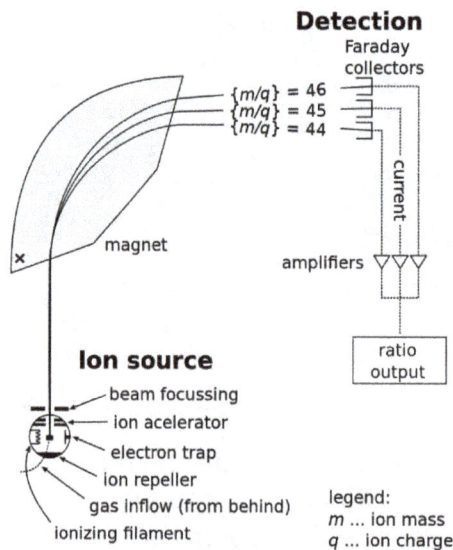

Schematics of a simple mass spectrometer with sector type mass analyzer. This one is for the measurement of carbon dioxide isotope ratios (IRMS) as in the carbon-13 urea breath test

A mass spectrometer consists of three components: an ion source, a mass analyzer, and a detector. The *ionizer* converts a portion of the sample into ions. There is a wide variety of ionization techniques, depending on the phase (solid, liquid, gas) of the sample and the efficiency of various ionization mechanisms for the unknown species. An extraction system removes ions from the sample, which are then targeted through the mass analyzer and onto the *detector*. The differences in masses of the fragments allows the mass analyzer to sort the ions by their mass-to-charge ratio. The detector measures the value of an indicator quantity and thus provides data for calculating the abundances of each ion present. Some detectors also give spatial information, e.g., a multichannel plate.

Theoretical Example

The following example describes the operation of a spectrometer mass analyzer, which is of the sector type. (Other analyzer types are treated below.) Consider a sample of sodium chloride (table salt). In the ion source, the sample is vaporized (turned into gas) and ionized (transformed into electrically charged particles) into sodium (Na^+) and chloride (Cl^-) ions. Sodium atoms and ions are monoisotopic, with a mass of about 23 u. Chloride atoms and ions come in two isotopes with masses of approximately 35 u (at a natural abundance of about 75 percent) and approximately 37 u (at a natural abundance of about 25 percent). The analyzer part of the spectrometer contains electric and magnetic fields, which exert forces on ions traveling through these fields. The speed of a charged particle may be increased or decreased while passing through the electric field, and its direction may be altered by the magnetic field. The magnitude of the deflection of the moving ion's trajectory depends on its mass-to-charge ratio. Lighter ions get deflected by the magnetic force more than heavier ions (based on Newton's second law of motion, $F = ma$). The streams of sorted ions pass from the analyzer to the detector, which records the relative abundance of each ion type. This information is used to determine the chemical element composition of the original sample (i.e. that both sodium and chlorine are present in the sample) and the isotopic composition of its constituents (the ratio of ^{35}Cl to ^{37}Cl).

Creating Ions

Surface ionization source at the Argonne National Laboratory linear accelerator

The ion source is the part of the mass spectrometer that ionizes the material under analysis (the analyte). The ions are then transported by magnetic or electric fields to the mass analyzer.

Techniques for ionization have been key to determining what types of samples can be analyzed by mass spectrometry. Electron ionization and chemical ionization are used for gases and vapors. In chemical ionization sources, the analyte is ionized by chemical ion-molecule reactions during collisions in the source. Two techniques often used with liquid and solid biological samples include electrospray ionization (invented by John Fenn) and matrix-assisted laser desorption/ionization (MALDI, initially developed as a similar technique "Soft Laser Desorption (SLD)" by K. Tanaka for which a Nobel Prize was awarded and as MALDI by M. Karas and F. Hillenkamp).

Hard Ionization and Soft Ionization

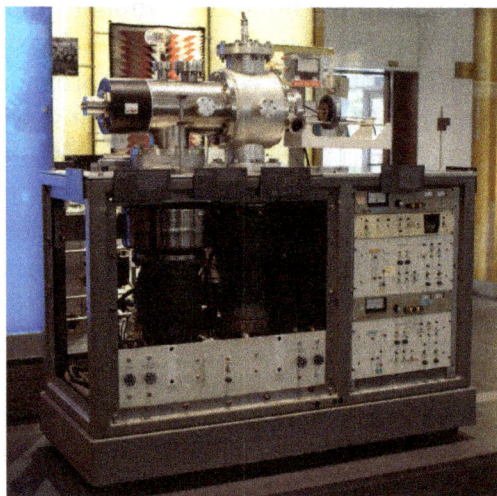

Quadrupole mass spectrometer and electrospray ion source used for Fenn's early work

In mass spectrometry, ionization refers to the production of gas phase ions suitable for resolution in the mass analyser or mass filter. Ionization occurs in the ion source. There are several ion sources available; each has advantages and disadvantages for particular applications. For example, electron ionization (EI) gives a high degree of fragmentation, yielding highly detailed mass spectra which when skilfully analysed can provide important information for structural elucidation/characterisation and facilitate identification of unknown compounds by comparison to mass spectral libraries obtained under identical operating conditions. However, EI is not suitable for coupling to HPLC, i.e. LC-MS, since at atmospheric pressure, the filaments used to generate electrons burn out rapidly. Thus EI is coupled predominantly with GC, i.e. GC-MS, where the entire system is under high vacuum.

Hard ionization techniques are processes which impart high quantities of residual energy in the subject molecule invoking large degrees of fragmentation (i.e. the systematic rupturing of bonds acts to remove the excess energy, restoring stability to the resulting ion). Resultant ions tend to have m/z lower than the molecular mass (other than in the case of proton transfer and not including isotope peaks). The most common example of hard ionization is electron ionization (EI).

Soft ionization refers to the processes which impart little residual energy onto the subject molecule and as such result in little fragmentation. Examples include fast atom bombardment (FAB), chemical ionization (CI), atmospheric-pressure chemical ionization (APCI), electrospray ionization (ESI), and matrix-assisted laser desorption/ionization (MALDI)

Inductively Coupled Plasma

Inductively coupled plasma ion source

Inductively coupled plasma (ICP) sources are used primarily for cation analysis of a wide array of sample types. In this source, a plasma that is electrically neutral overall, but that has had a substantial fraction of its atoms ionized by high temperature, is used to atomize introduced sample molecules and to further strip the outer electrons from those atoms. The plasma is usually generated from argon gas, since the first ionization energy of argon atoms is higher than the first of any other elements except He, O, F and Ne, but lower than the second ionization energy of all except the most electropositive metals. The heating is achieved by a radio-frequency current passed through a coil surrounding the plasma.

Other Ionization Techniques

Others include photoionization, glow discharge, field desorption (FD), fast atom bombardment (FAB), thermospray, desorption/ionization on silicon (DIOS), Direct Analysis in Real Time (DART), atmospheric pressure chemical ionization (APCI), secondary ion mass spectrometry (SIMS), spark ionization and thermal ionization (TIMS).

Mass Selection

Mass analyzers separate the ions according to their mass-to-charge ratio. The following two laws govern the dynamics of charged particles in electric and magnetic fields in vacuum:

$\mathbf{F} = Q(\mathbf{E} + \mathbf{v} \times \mathbf{B})$ (Lorentz force law);

$\mathbf{F} = m\mathbf{a}$ (Newton's second law of motion in non-relativistic case, i.e. valid only at ion velocity much lower than the speed of light).

Here F is the force applied to the ion, m is the mass of the ion, a is the acceleration, Q is the ion charge, E is the electric field, and v × B is the vector cross product of the ion velocity and the magnetic field

Equating the above expressions for the force applied to the ion yields:

$(m/Q)\mathbf{a} = \mathbf{E} + \mathbf{v} \times \mathbf{B}$.

This differential equation is the classic equation of motion for charged particles. Together with the particle's initial conditions, it completely determines the particle's motion in space and time in terms of m/Q. Thus mass spectrometers could be thought of as "mass-to-charge spectrometers". When presenting data, it is common to use the (officially) dimensionless m/z, where z is the number of elementary charges (e) on the ion (z=Q/e). This quantity, although it is informally called the mass-to-charge ratio, more accurately speaking represents the ratio of the mass number and the charge number, z.

There are many types of mass analyzers, using either static or dynamic fields, and magnetic or electric fields, but all operate according to the above differential equation. Each analyzer type has its strengths and weaknesses. Many mass spectrometers use two or more mass analyzers for tandem mass spectrometry (MS/MS). In addition to the more common mass analyzers listed below, there are others designed for special situations.

There are several important analyser characteristics. The mass resolving power is the measure of the ability to distinguish two peaks of slightly different m/z. The mass accuracy is the ratio of the m/z measurement error to the true m/z. Mass accuracy is usually measured in ppm or milli mass units. The mass range is the range of m/z amenable to analysis by a given analyzer. The linear dynamic range is the range over which ion signal is linear with analyte concentration. Speed refers to the time frame of the experiment and ultimately is used to determine the number of spectra per unit time that can be generated.

Sector Instruments

A sector field mass analyzer uses an electric and/or magnetic field to affect the path and/or velocity of the charged particles in some way. As shown above, sector instruments bend the trajectories of the ions as they pass through the mass analyzer, according to their mass-to-charge ratios, deflecting the more charged and faster-moving, lighter ions more. The analyzer can be used to select a narrow range of m/z or to scan through a range of m/z to catalog the ions present.

ThermoQuest AvantGarde sector mass spectrometer

Time-of-flight

The time-of-flight (TOF) analyzer uses an electric field to accelerate the ions through the same potential, and then measures the time they take to reach the detector. If the particles all have the same charge, the kinetic energies will be identical, and their velocities will depend only on their masses. Ions with a lower mass will reach the detector first.

Quadrupole Mass Filter

Quadrupole mass analyzers use oscillating electrical fields to selectively stabilize or destabilize the paths of ions passing through a radio frequency (RF) quadrupole field created between 4 parallel rods. Only the ions in a certain range of mass/charge ratio are passed through the system at any time, but changes to the potentials on the rods allow a wide range of m/z values to be swept rapidly, either continuously or in a succession of discrete hops. A quadrupole mass analyzer acts as a mass-selective filter and is closely related to the quadrupole ion trap, particularly the linear quadrupole ion trap except that it is designed to pass the untrapped ions rather than collect the trapped ones, and is for that reason referred to as a transmission quadrupole. A magnetically enhanced quadrupole mass analyzer includes the addition of a magnetic field, either applied axially or transversely. This novel type of instrument leads to an additional performance enhancement in terms of resolution and/or sensitivity depending upon the magnitude and orientation of the applied magnetic field. A common variation of the transmission quadrupole is the triple quadrupole mass spectrometer. The "triple quad" has three consecutive quadrupole stages, the first acting as a mass filter to transmit a particular incoming ion to the second quadrupole, a collision chamber, wherein that ion can be broken into fragments. The third quadrupole also acts as a mass filter, to transmit a particular fragment ion to the detector. If a quadrupole is made to rapidly and repetitively cycle through a range of mass filter settings, full spectra can be reported. Likewise, a triple quad can be made to perform various scan types characteristic of tandem mass spectrometry.

Ion Traps

Three-dimensional Quadrupole Ion Trap

The quadrupole ion trap works on the same physical principles as the quadrupole mass analyzer, but the ions are trapped and sequentially ejected. Ions are trapped in a mainly quadrupole RF field, in a space defined by a ring electrode (usually connected to the main RF potential) between two endcap electrodes (typically connected to DC or auxiliary AC potentials). The sample is ionized either internally (e.g. with an electron or laser beam), or externally, in which case the ions are often introduced through an aperture in an endcap electrode.

There are many mass/charge separation and isolation methods but the most commonly used is the mass instability mode in which the RF potential is ramped so that the orbit of ions with a mass $a > b$ are stable while ions with mass b become unstable and are ejected on the z-axis onto a detector. There are also non-destructive analysis methods.

Ions may also be ejected by the resonance excitation method, whereby a supplemental oscillatory excitation voltage is applied to the endcap electrodes, and the trapping voltage amplitude and/or excitation voltage frequency is varied to bring ions into a resonance condition in order of their mass/charge ratio.

Cylindrical Ion Trap

The cylindrical ion trap mass spectrometer (CIT) is a derivative of the quadrupole ion trap where the electrodes are formed from flat rings rather than hyperbolic shaped electrodes. The architecture lends itself well to miniaturization because as the size of a trap is reduced, the shape of the electric field near the center of the trap, the region where the ions are trapped, forms a shape similar to that of a hyperbolic trap.

Linear Quadrupole Ion Trap

A linear quadrupole ion trap is similar to a quadrupole ion trap, but it traps ions in a two dimensional quadrupole field, instead of a three-dimensional quadrupole field as in a 3D quadrupole ion trap. Thermo Fisher's LTQ ("linear trap quadrupole") is an example of the linear ion trap.

A toroidal ion trap can be visualized as a linear quadrupole curved around and connected at the ends or as a cross section of a 3D ion trap rotated on edge to form the toroid, donut shaped trap. The trap can store large volumes of ions by distributing them throughout the ring-like trap structure. This toroidal shaped trap is a configuration that allows the increased miniaturization of an ion trap mass analyzer. Additionally all ions are stored in the same trapping field and ejected together simplifying detection that can be complicated with array configurations due to variations in detector alignment and machining of the arrays.

As with the toroidal trap, linear traps and 3D quadrupole ion traps are the most commonly miniaturized mass analyzers due to their high sensitivity, tolerance for mTorr pressure, and capabilities for single analyzer tandem mass spectrometry (e.g. product ion scans).

Orbitrap

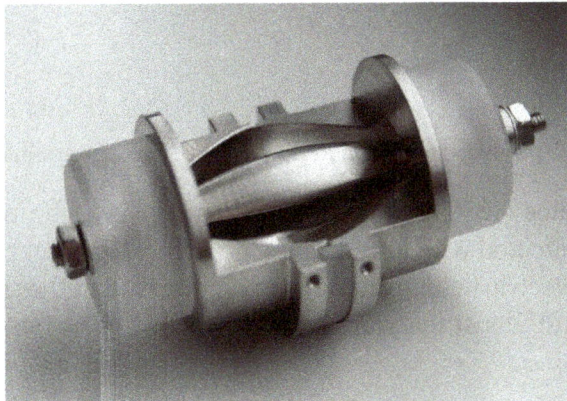

Orbitrap mass analyzer

Orbitrap instruments are similar to Fourier transform ion cyclotron resonance mass spectrometers. Ions are electrostatically trapped in an orbit around a central, spindle shaped electrode. The electrode confines the ions so that they both orbit around the central electrode and oscillate back and forth along the central electrode's long axis. This oscillation generates an image current in the detector plates which is recorded by the instrument. The frequencies of these image currents depend on the mass to charge ratios of the ions. Mass spectra are obtained by Fourier transformation of the recorded image currents.

Orbitraps have a high mass accuracy, high sensitivity and a good dynamic range.

Fourier Transform Ion Cyclotron Resonance

Fourier transform mass spectrometry (FTMS), or more precisely Fourier transform ion cyclotron resonance MS, measures mass by detecting the image current produced by ions cyclotroning in the presence of a magnetic field. Instead of measuring the deflection of ions with a detector such as an electron multiplier, the ions are injected into a Penning trap (a static electric/magnetic ion trap) where they effectively form part of a circuit. Detectors at fixed positions in space measure the electrical signal of ions which pass near them over time, producing a periodic signal. Since the frequency of an ion's cycling is determined by its mass to charge ratio, this can be deconvoluted by performing a Fourier transform on the signal. FTMS has the advantage of high sensitivity (since each ion is "counted" more than once) and much higher resolution and thus precision.

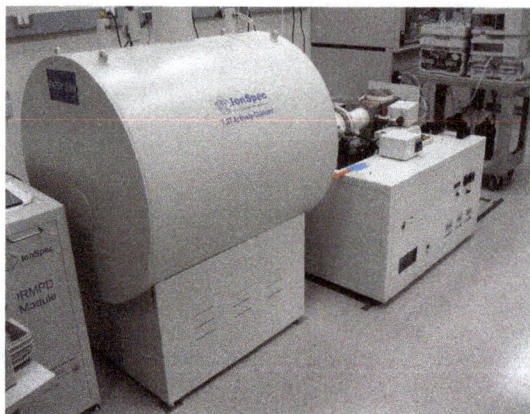

Fourier transform ion cyclotron resonance mass spectrometer

Ion cyclotron resonance (ICR) is an older mass analysis technique similar to FTMS except that ions are detected with a traditional detector. Ions trapped in a Penning trap are excited by an RF electric field until they impact the wall of the trap, where the detector is located. Ions of different mass are resolved according to impact time.

Detectors

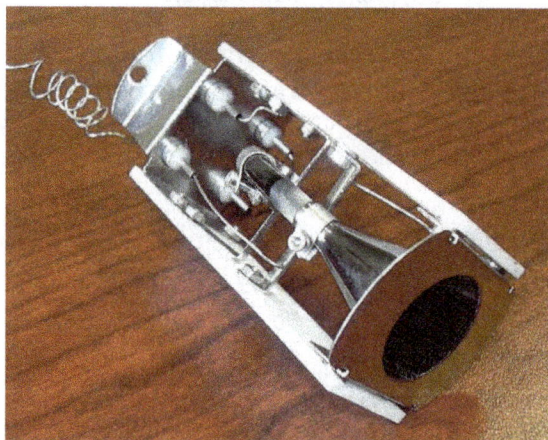

A continuous dynode particle multiplier detector.

The final element of the mass spectrometer is the detector. The detector records either the charge induced or the current produced when an ion passes by or hits a surface. In a scanning instrument, the signal produced in the detector during the course of the scan versus where the instrument is in the scan (at what m/Q) will produce a mass spectrum, a record of ions as a function of m/Q.

Typically, some type of electron multiplier is used, though other detectors including

Faraday cups and ion-to-photon detectors are also used. Because the number of ions leaving the mass analyzer at a particular instant is typically quite small, considerable amplification is often necessary to get a signal. Microchannel plate detectors are commonly used in modern commercial instruments. In FTMS and Orbitraps, the detector consists of a pair of metal surfaces within the mass analyzer/ion trap region which the ions only pass near as they oscillate. No direct current is produced, only a weak AC image current is produced in a circuit between the electrodes. Other inductive detectors have also been used.

Tandem Mass Spectrometry

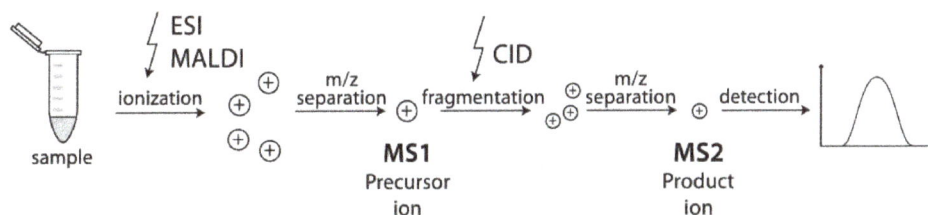

Tandem mass spectrometry for biological molecules using ESI or MALDI

A tandem mass spectrometer is one capable of multiple rounds of mass spectrometry, usually separated by some form of molecule fragmentation. For example, one mass analyzer can isolate one peptide from many entering a mass spectrometer. A second mass analyzer then stabilizes the peptide ions while they collide with a gas, causing them to fragment by collision-induced dissociation (CID). A third mass analyzer then sorts the fragments produced from the peptides. Tandem MS can also be done in a single mass analyzer over time, as in a quadrupole ion trap. There are various methods for fragmenting molecules for tandem MS, including collision-induced dissociation (CID), electron capture dissociation (ECD), electron transfer dissociation (ETD), infrared multiphoton dissociation (IRMPD), blackbody infrared radiative dissociation (BIRD), electron-detachment dissociation (EDD) and surface-induced dissociation (SID). An important application using tandem mass spectrometry is in protein identification.

Tandem mass spectrometry enables a variety of experimental sequences. Many commercial mass spectrometers are designed to expedite the execution of such routine sequences as selected reaction monitoring (SRM) and precursor ion scanning. In SRM, the first analyzer allows only a single mass through and the second analyzer monitors for multiple user-defined fragment ions. SRM is most often used with scanning instruments where the second mass analysis event is duty cycle limited. These experiments are used to increase specificity of detection of known molecules, notably in pharmacokinetic studies. Precursor ion scanning refers to monitoring for a specific loss from the precursor ion. The first and second mass analyzers scan across the spectrum as partitioned by a user-defined m/z value. This experiment is used to detect specific motifs within unknown molecules.

Another type of tandem mass spectrometry used for radiocarbon dating is accelerator mass spectrometry (AMS), which uses very high voltages, usually in the mega-volt range, to accelerate negative ions into a type of tandem mass spectrometer.

Common Mass Spectrometer Configurations and Techniques

When a specific configuration of source, analyzer, and detector becomes conventional in practice, often a compound acronym arises to designate it, and the compound acronym may be better known among nonspectrometrists than the component acronyms. The epitome of this is MALDI-TOF, which simply refers to combining a matrix-assisted laser desorption/ionization source with a time-of-flight mass analyzer. The MALDI-TOF moniker is more widely recognized by the non-mass spectrometrists than MALDI or TOF individually. Other examples include inductively coupled plasma-mass spectrometry (ICP-MS), accelerator mass spectrometry (AMS), thermal ionization-mass spectrometry (TIMS) and spark source mass spectrometry (SSMS). Sometimes the use of the generic "MS" actually connotes a very specific mass analyzer and detection system, as is the case with AMS, which is always sector based.

Certain applications of mass spectrometry have developed monikers that although strictly speaking would seem to refer to a broad application, in practice have come instead to connote a specific or a limited number of instrument configurations. An example of this is isotope ratio mass spectrometry (IRMS), which refers in practice to the use of a limited number of sector based mass analyzers; this name is used to refer to both the application and the instrument used for the application.

Separation Techniques Combined with Mass Spectrometry

An important enhancement to the mass resolving and mass determining capabilities of mass spectrometry is using it in tandem with chromatographic and other separation techniques.

Gas Chromatography

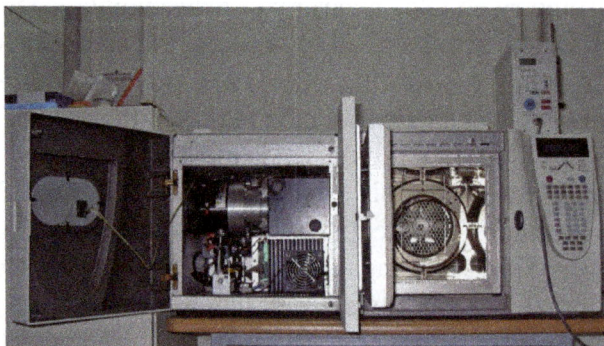

A gas chromatograph (right) directly coupled to a mass spectrometer (left)

A common combination is gas chromatography-mass spectrometry (GC/MS or GC-MS). In this technique, a gas chromatograph is used to separate different compounds. This stream of separated compounds is fed online into the ion source, a metallic filament to which voltage is applied. This filament emits electrons which ionize the compounds. The ions can then further fragment, yielding predictable patterns. Intact ions and fragments pass into the mass spectrometer's analyzer and are eventually detected.

Liquid Chromatography

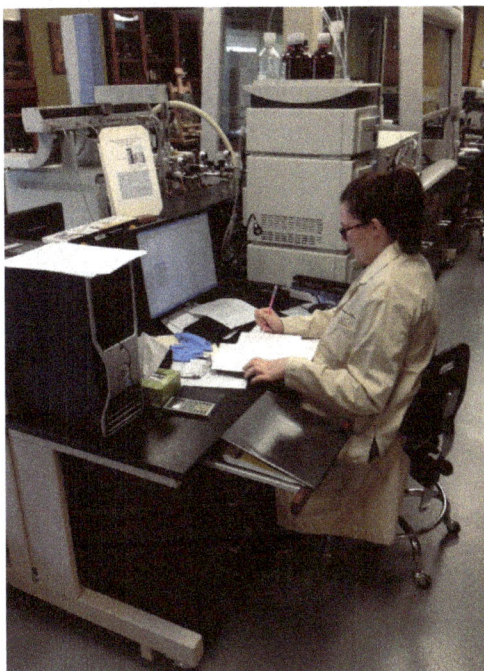

Indianapolis Museum of Art conservation scientist performing
liquid chromatography–mass spectrometry.

Similar to gas chromatography MS (GC/MS), liquid chromatography-mass spectrometry (LC/MS or LC-MS) separates compounds chromatographically before they are introduced to the ion source and mass spectrometer. It differs from GC/MS in that the mobile phase is liquid, usually a mixture of water and organic solvents, instead of gas. Most commonly, an electrospray ionization source is used in LC/MS. Other popular and commercially available LC/MS ion sources are atmospheric pressure chemical ionization and atmospheric pressure photoionization. There are also some newly developed ionization techniques like laser spray.

Capillary Electrophoresis–mass Spectrometry

Capillary electrophoresis–mass spectrometry (CE-MS) is a technique that combines the liquid separation process of capillary electrophoresis with mass spectrometry. CE-MS is typically coupled to electrospray ionization.

Ion Mobility

Ion mobility spectrometry-mass spectrometry (IMS/MS or IMMS) is a technique where ions are first separated by drift time through some neutral gas under an applied electrical potential gradient before being introduced into a mass spectrometer. Drift time is a measure of the radius relative to the charge of the ion. The duty cycle of IMS (the time over which the experiment takes place) is longer than most mass spectrometric techniques, such that the mass spectrometer can sample along the course of the IMS separation. This produces data about the IMS separation and the mass-to-charge ratio of the ions in a manner similar to LC/MS.

The duty cycle of IMS is short relative to liquid chromatography or gas chromatography separations and can thus be coupled to such techniques, producing triple modalities such as LC/IMS/MS.

Data and Analysis

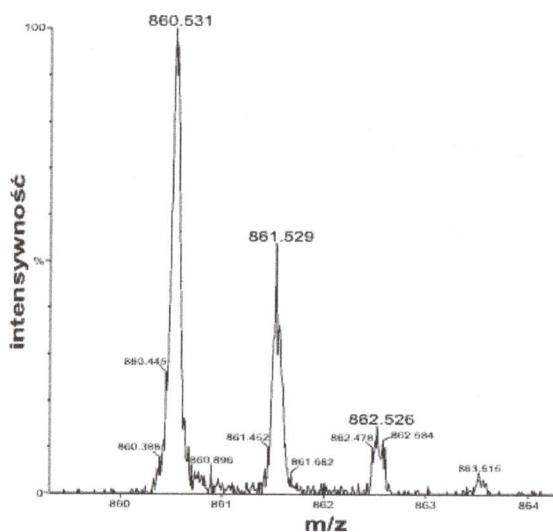

Mass spectrum of a peptide showing the isotopic distribution

Data Representations

Mass spectrometry produces various types of data. The most common data representation is the mass spectrum.

Certain types of mass spectrometry data are best represented as a mass chromatogram. Types of chromatograms include selected ion monitoring (SIM), total ion current (TIC), and selected reaction monitoring (SRM), among many others.

Other types of mass spectrometry data are well represented as a three-dimensional contour map. In this form, the mass-to-charge, m/z is on the x-axis, intensity the y-axis, and an additional experimental parameter, such as time, is recorded on the z-axis.

Data Analysis

Mass spectrometry data analysis is specific to the type of experiment producing the data. General subdivisions of data are fundamental to understanding any data.

Many mass spectrometers work in either *negative ion mode* or *positive ion mode*. It is very important to know whether the observed ions are negatively or positively charged. This is often important in determining the neutral mass but it also indicates something about the nature of the molecules.

Different types of ion source result in different arrays of fragments produced from the original molecules. An electron ionization source produces many fragments and mostly single-charged (1-) radicals (odd number of electrons), whereas an electrospray source usually produces non-radical quasimolecular ions that are frequently multiply charged. Tandem mass spectrometry purposely produces fragment ions post-source and can drastically change the sort of data achieved by an experiment.

Knowledge of the origin of a sample can provide insight into the component molecules of the sample and their fragmentations. A sample from a synthesis/manufacturing process will probably contain impurities chemically related to the target component. A crudely prepared biological sample will probably contain a certain amount of salt, which may form adducts with the analyte molecules in certain analyses.

Results can also depend heavily on sample preparation and how it was run/introduced. An important example is the issue of which matrix is used for MALDI spotting, since much of the energetics of the desorption/ionization event is controlled by the matrix rather than the laser power. Sometimes samples are spiked with sodium or another ion-carrying species to produce adducts rather than a protonated species.

Mass spectrometry can measure molar mass, molecular structure, and sample purity. Each of these questions requires a different experimental procedure; therefore, adequate definition of the experimental goal is a prerequisite for collecting the proper data and successfully interpreting it.

Interpretation of Mass Spectra

Since the precise structure or peptide sequence of a molecule is deciphered through the set of fragment masses, the interpretation of mass spectra requires combined use of various techniques. Usually the first strategy for identifying an unknown compound is to compare its experimental mass spectrum against a library of mass spectra. If no matches result from the search, then manual interpretation or software assisted interpretation of mass spectra must be performed. Computer simulation of ionization and fragmentation processes occurring in mass spectrometer is the primary tool for assigning structure or peptide sequence to a molecule. An *a priori* structural information is fragmented *in silico* and the resulting pattern is compared with observed spectrum.

Such simulation is often supported by a fragmentation library that contains published patterns of known decomposition reactions. Software taking advantage of this idea has been developed for both small molecules and proteins.

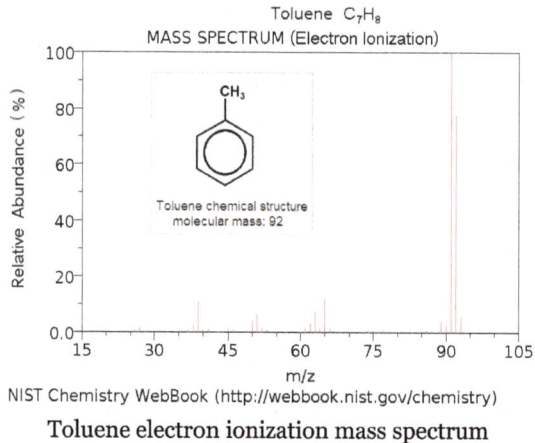

Toluene C$_7$H$_8$
MASS SPECTRUM (Electron Ionization)

Toluene chemical structure
molecular mass: 92

NIST Chemistry WebBook (http://webbook.nist.gov/chemistry)
Toluene electron ionization mass spectrum

Analysis of mass spectra can also be spectra with accurate mass. A mass-to-charge ratio value (m/z) with only integer precision can represent an immense number of theoretically possible ion structures; however, more precise mass figures significantly reduce the number of candidate molecular formulas. A computer algorithm called formula generator calculates all molecular formulas that theoretically fit a given mass with specified tolerance.

A recent technique for structure elucidation in mass spectrometry, called precursor ion fingerprinting, identifies individual pieces of structural information by conducting a search of the tandem spectra of the molecule under investigation against a library of the product-ion spectra of structurally characterized precursor ions.

Applications

NOAA Particle Analysis by Laser Mass Spectrometry aerosol mass spectrometer
aboard a NASA WB-57 high-altitude research aircraft

Mass spectrometry has both qualitative and quantitative uses. These include identifying unknown compounds, determining the isotopic composition of elements in a molecule, and determining the structure of a compound by observing its fragmentation. Other uses include quantifying the amount of a compound in a sample or studying the fundamentals of gas phase ion chemistry (the chemistry of ions and neutrals in a vacuum). MS is now in very common use in analytical laboratories that study physical, chemical, or biological properties of a great variety of compounds.

As an analytical technique it possesses distinct advantages such as: Increased sensitivity over most other analytical techniques because the analyzer, as a mass-charge filter, reduces background interference, Excellent specificity from characteristic fragmentation patterns to identify unknowns or confirm the presence of suspected compounds, Information about molecular weight, Information about the isotopic abundance of elements, Temporally resolved chemical data.

A few of the disadvantages of the method is that often fails to distinguish between optical and geometrical isomers and the positions of substituent in o-, m- and p- positions in an aromatic ring. Also, its scope is limited in identifying hydrocarbons that produce similar fragmented ions.

Isotope Ratio MS: Isotope Dating and Tracing

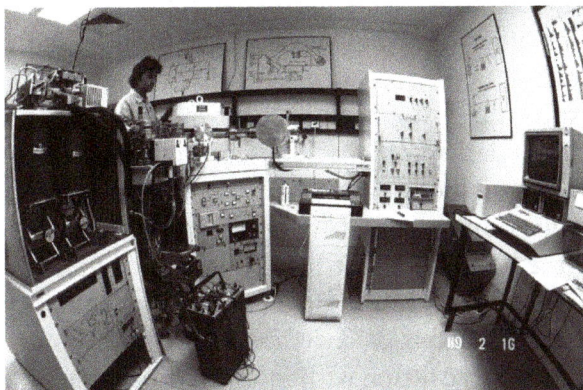

Mass spectrometer to determine the $^{16}O/^{18}O$ and $^{12}C/^{13}C$ isotope ratio on biogenous carbonate

Mass spectrometry is also used to determine the isotopic composition of elements within a sample. Differences in mass among isotopes of an element are very small, and the less abundant isotopes of an element are typically very rare, so a very sensitive instrument is required. These instruments, sometimes referred to as isotope ratio mass spectrometers (IR-MS), usually use a single magnet to bend a beam of ionized particles towards a series of Faraday cups which convert particle impacts to electric current. A fast on-line analysis of deuterium content of water can be done using flowing afterglow mass spectrometry, FA-MS. Probably the most sensitive and accurate mass spectrometer for this purpose is the accelerator mass spectrometer (AMS). This is because it provides ultimate sensitivity, capable of measuring individual atoms and measuring nuclides with a dynamic range of ~10^{15} relative to the major stable isotope. Isotope ratios are important markers of a variety of processes. Some isotope ratios are used to determine the age of materials for example as in carbon dating. Labeling with stable isotopes is also used for protein quantification.

Trace Gas Analysis

Several techniques use ions created in a dedicated ion source injected into a flow tube or a drift tube: selected ion flow tube (SIFT-MS), and proton transfer reaction (PTR-MS), are variants of chemical ionization dedicated for trace gas analysis of air, breath or liquid headspace using well defined reaction time allowing calculations of analyte concentrations from the known reaction kinetics without the need for internal standard or calibration.

Atom Probe

An atom probe is an instrument that combines time-of-flight mass spectrometry and field-evaporation microscopy to map the location of individual atoms.

Pharmacokinetics

Pharmacokinetics is often studied using mass spectrometry because of the complex nature of the matrix (often blood or urine) and the need for high sensitivity to observe low dose and long time point data. The most common instrumentation used in this application is LC-MS with a triple quadrupole mass spectrometer. Tandem mass spectrometry is usually employed for added specificity. Standard curves and internal standards are used for quantitation of usually a single pharmaceutical in the samples. The samples represent different time points as a pharmaceutical is administered and then metabolized or cleared from the body. Blank or t=0 samples taken before administration are important in determining background and ensuring data integrity with such complex sample matrices. Much attention is paid to the linearity of the standard curve; however it is not uncommon to use curve fitting with more complex functions such as quadratics since the response of most mass spectrometers is less than linear across large concentration ranges.

There is currently considerable interest in the use of very high sensitivity mass spectrometry for microdosing studies, which are seen as a promising alternative to animal experimentation.

Protein Characterization

Mass spectrometry is an important method for the characterization and sequencing of proteins. The two primary methods for ionization of whole proteins are electrospray ionization (ESI) and matrix-assisted laser desorption/ionization (MALDI). In keeping with the performance and mass range of available mass spectrometers, two approaches are used for characterizing proteins. In the first, intact proteins are ionized by either of the two techniques described above, and then introduced to a mass analyzer. This approach is referred to as "top-down" strategy of protein analysis. In the second, proteins are enzymatically digested into smaller peptides using proteases such as trypsin or pepsin, either in solution or in gel after electrophoretic separation. Other proteolytic agents are also used. The collection of peptide products are then introduced to the mass analyzer. When the characteristic pattern of peptides is used for the identification of the protein the method is called peptide mass fingerprinting (PMF), if the identification is performed using the sequence data determined in tandem MS analysis it is called de novo peptide sequencing. These procedures of protein analysis are also referred to as the "bottom-up" approach.

Glycan Analysis

Mass spectrometry (MS), with its low sample requirement and high sensitivity, has been predominantly used in glycobiology for characterization and elucidation of glycan structures. Mass spectrometry provides a complementary method to HPLC for the analysis of glycans. Intact glycans may be detected directly as singly charged ions by matrix-assisted laser desorption/ionization mass spectrometry (MALDI-MS) or, following permethylation or peracetylation, by fast atom bombardment mass spectrometry (FAB-MS). Electrospray ionization mass spectrometry (ESI-MS) also gives good signals for the smaller glycans. Various free and commercial software are now available which interpret MS data and aid in Glycan structure characterization.

Space Exploration

NASA's Phoenix Mars Lander analyzing a soil sample from the "Rosy Red" trench with the TEGA mass spectrometer

As a standard method for analysis, mass spectrometers have reached other planets and moons. Two were taken to Mars by the Viking program. In early 2005 the Cassini–Huygens mission delivered a specialized GC-MS instrument aboard the Huygens probe through the atmosphere of Titan, the largest moon of the planet Saturn. This instrument analyzed atmospheric samples along its descent trajectory and was able to vaporize and analyze samples of Titan's frozen, hydrocarbon covered surface once the probe had landed. These measurements compare the abundance of isotope(s) of each particle comparatively to earth's natural abundance. Also on board the Cassini–Huygens spacecraft is an ion and neutral mass spectrometer which has been taking measurements of Titan's atmospheric composition as well as the composition of Enceladus' plumes. A Thermal and Evolved Gas Analyzer mass spectrometer was carried by the Mars Phoenix Lander launched in 2007.

Mass spectrometers are also widely used in space missions to measure the composition of plasmas. For example, the Cassini spacecraft carries the Cassini Plasma Spectrometer (CAPS), which measures the mass of ions in Saturn's magnetosphere.

Respired Gas Monitor

Mass spectrometers were used in hospitals for respiratory gas analysis beginning around 1975 through the end of the century. Some are probably still in use but none are currently being manufactured.

Found mostly in the operating room, they were a part of a complex system, in which respired gas samples from patients undergoing anesthesia were drawn into the instrument through a valve mechanism designed to sequentially connect up to 32 rooms to the mass spectrometer. A computer directed all operations of the system. The data collected from the mass spectrometer was delivered to the individual rooms for the anesthesiologist to use.

The uniqueness of this magnetic sector mass spectrometer may have been the fact that a plane of detectors, each purposely positioned to collect all of the ion species expected to be in the samples, allowed the instrument to simultaneously report all of the gases respired by the patient. Although the mass range was limited to slightly over 120 u, fragmentation of some of the heavier molecules negated the need for a higher detection limit.

Preparative Mass Spectrometry

The primary function of mass spectrometry is as a tool for chemical analyses based on detection and quantification of ions according to their mass-to-charge ratio. However, mass spectrometry also shows promise for material synthesis. Ion soft landing is characterized by deposition of intact species on surfaces at low kinetic energies which precludes the fragmentation of the incident species. The soft landing technique was first reported in 1977 for the reaction of low energy sulfur containing ions on a lead surface.

Chemcatcher

Chemcatcher deployment kit

Chemcatcher® is a highly versatile and cost-effective passive sampling device for monitoring a wide variety of pollutants in water. Most monitoring programmes involve the periodic collection of low volume spot samples (bottle or grab) of water, which is challenging, particularly where levels fluctuate over time and when chemicals are only present at trace, yet toxicologically relevant concentrations. The Chemcatcher® passive sampling device is currently being used throughout the world to measure time-weighted average (TWA) or equilibrium concentrations of a wide range of pollutants in water. This allows the end user to obtain a more representative picture of the chemicals that may be present in the aquatic environment. The Chemcatcher® concept was developed by Professors Richard Greenwood and Graham Mills at the University of Portsmouth, together with colleagues from Chalmers University of Technology, Sweden. The device is patented in a number of countries and the name is a registered trademark in the United Kingdom.

Chemcatcher® comprises a robust, reusable three-component a low-cost, three component, water-tight PTFE body. Two different designs are available to accommodate different types of commercially available 47 mm diameter receiving phase disks:

Chemcatcher kit

- 3M Empore™ Chemcatcher®: weighs 83 g and houses the 47 mm 3M Empore™ range of receiving phases (e.g. C18, SDB, chelating, anion and cation exchange and RAD disks) . As well as the appropriate Empore™ receiving disk, you will need a diffusion-limiting membrane to prepare the device for deployment.

- Horizon Atlantic™ Chemcatcher®: weighs 130 g and houses the 47 mm Horizon Atlantic™ SPE disks as receiving phases (e.g. HLB, C18 and DVB). As well as the appropriate Atlantic™ receiving disk, you will need a diffusion-limiting membrane to prepare the device for deployment.

The use of quality-controlled, commercially available receiving phases allow high reproducibility as compared to some other passive sampling devices. The bound receiving phase also ensures that the active sampling area of the device remains constant during laboratory or field calibration uptake experiments and in field deployments. By altering the combination of receiving phase and diffusion-limiting membrane (low-density polyethylene, polyethersulphone or cellulose acetate), you can monitor potable, surface , coastal and marine environment. The sampler can be deployed in the field for extended periods of time ranging from days to weeks. The specific pollutants of interest are sequestered by the samplers and these are retained on the receiving phase disk. After retrieval from the environment the pollutants are eluted from the disk and analysed in the laboratory using conventional instrumental methods. In order to obtain TWA concentrations the sampler must first be calibrated in the laboratory so as to ascertain the uptake rate (usually measured as the volume of water cleared per unit time i.e. L/h for the analyte) of the pollutant of interest. The Chemcatcher® has been used in a range of aquatic environments; however, most work to date has been in monitoring the TWA concentrations of priority and emerging pollutants surface waters.

The use of passive sampling devices, such as the Chemcatcher®, and the polar organic chemical integrative sampler (POCIS), have a number of advantages over the use of spot or bottle sampling for monitoring pollutants in the aquatic environment. The latter technique gives only an instantaneous concentration of the pollutant as the specific time of sampling. Passive samplers, depending on their mode of use can give either the TWA or equilibrium concentration of the pollutant over the deployment period. The measurement of TWA concentrations give a better indication of the long-term environmental conditions and enables improved risk assessment. Such devices have potential roles in monitoring programmes such as those underpinning the European Union's Water Framework Directive and Marine Strategy Framework Directive. An ISO standard concerning the use of passive samplers for the determination of priority pollutants in surface waters is available for end users of this technology.

Polar Organic Chemical Integrative Sampler

A polar organic chemical integrative sampler (POCIS) is a passive sampling device which allows for the in situ collection of a time-integrated average of hydrophilic organic contaminants developed by researchers with the United States Geological Survey in Columbia, Missouri. POCIS provides a means for estimating the toxicological significance of waterborne contaminants. The POCIS sampler mimics the respiratory exposure of organisms living in the aquatic environment and can provide an understanding of bioavailable contaminants present in the system. POCIS can be deployed in a wide range of aquatic environments and is commonly used to assist in environmental monitoring studies.

Background

The first passive sampling devices were developed in the 1970s to determine concentrations of contaminants in the air. In 1980 this technology was first adapted for the monitoring of organic contaminants in water. The initial type of passive sampler developed for aquatic monitoring purposes was the semipermeable membrane device (SMPD). SPMD samplers are most effective at absorbing hydrophobic pollutants with a log octanol-water partitioning coefficient (Kow) ranging from 4-8. As the global emission of bioconcentratable persistent organic pollutants (POPs) was shown to result in adverse ecological effects, industry developed a wide range of increasing water-soluble, polar hydrophilic organic compounds (HpOCs) to replace them. These compounds generally have lower bioconcentration factors. However, there is evidence that large fluxes of these HpOCs into aquatic environments may be responsible for a number of adverse effects to aquatic organisms, such as altered behavior, neurotoxicity, endocrine disruption, and impaired reproduction. In the late 1990s research was underway to develop a new passive sampler in order to monitor HpOCs with a log Kow value of less than 3. In 1999 the POCIS sampler was under development at the University of Missouri-Columbia. It gathered more support in the early 2000s as concern increased regarding the effects of pharmaceutical and personal care products in surface waters.

The United States Geological Survey (USGS) has been heavily involved in the development of passive samplers and has articles in their database regarding the development of POCIS as early as 2000. The USGS Columbia Environmental Research Center (CERC) is a self-proclaimed international leader in the field of passive sampling. There have been recent efforts by the USGS to connect people who have an interest in passive sampling. An international workshop and symposium on passive sampling was held by the USGS in 2013 to connect developers, policy makers and end users in order to discuss ways of monitoring environmental pollution.

Fundamentals

The POCIS device was developed and patented by Jimmie D. Petty, James N. Huckins, and David A. Alvarez, of the Columbia Environmental Research Center. Integrative passive samplers are an effective way to monitor the concentration of organic contaminants in aquatic systems over time. Most aquatic monitoring programs rely on collecting individual samples, often called grab samples, at a specific time. The grab sampling method is associated with many disadvantages that can be resolved by passive sampling techniques. When contaminants are present in trace amounts, grab sampling may require the collection of large volumes of water. Also, lab analysis of the sample can only provide a snapshot of contaminant levels at the time of collection. This approach therefore has drawbacks when monitoring in environments where water contamination varies over time and episodic contamination events occur. Passive sampling techniques have been able to provide a time-integrated sample of water contamination with low detection limits and in situ extraction of analytes.

POCIS Set-up

The POCIS sampler consists of an array of sampling disks mounted on a support rod. Each disk consists of a solid sorbent sandwiched between two polyethersoulfone (PES) microporous membranes which are then compressed between two stainless steel rings which expose a sampling area. A standard POCIS disk consists of a sampling surface area to sorbent mass ratio of approximately 180 cm^2g. Because the amount of chemical sampled is directly related to the sample surface area, it is sometimes necessary to combine extracts from multiple POCIS disks into one sample. Stainless steel rings, or other rigid inert material, are essential to prevent sorbent loss as the PES membranes are not able to be heat sealed. The POCIS array is then inserted and deployed within a protective canister. This canister is usually made of stainless steel or PVC and works to deflect debris that may displace the POCIS array during its deployment.

The PES membrane acts as a semipermeable barrier between the sorbent and surrounding aquatic environment. It allows dissolved contaminants to pass through the sorbent while selectively excluding any particles larger than 100 nm. The membrane resists biofouling because the polyethersulphone used in the design is less prone than other materials. The POCIS is versatile in that the sorbents can be changed to target different classes of contaminants. However, only two sorbent classes are considered as standards of all POCIS deployments to date.

Theory and Modeling

Each POCIS disk will sample a certain volume of water per day. The volume of water sampled varies from chemical to chemical and is dependent on the physical and chemical properties of the compound as well as the duration of sampling. The sampling rate of POCIS can vary with changes in the water flow, turbulence, temperature, and the buildup of solids on the sampler's surface. The accumulation of contaminants into a POCIS device is the result of three successive process occurring at the same time. First, the contaminants have to diffuse across the water boundary layer. The thickness of this layer is dependent on water flow and turbulence around the sampler and can significantly alter sampling rates. Second, the contaminant must transport across the membrane either through the water-filled pores or through the membrane itself. Finally, contaminants transfer from the membrane into the sorbent material mainly through adsorption. These last two steps make the modeling, understanding, and prediction of accumulation by a POCIS device challenging. To date, a limited number of chemical sampling rates have been determined.

Accumulation of chemicals by a POCIS device generally follows first order kinetics. The kinetics are characterized by an initial integrative phase, followed by an equilibrium partitioning phase. During the integrative phase of uptake, a passive sampling device accumulates residues linearly relative to time, assuming constant exposure concentrations. Based on current results, the POCIS sampler remains in a linear phase for at least 30 days, and has been observed up to 56 days. Therefore, both laboratory and field data justify the use of a linear uptake model for the calculation of sample rates. In order to estimate the ambient water concentration of contaminants sampled by a POCIS device, there must be available calibration data applicable for in situ conditions regarding the target compound. Currently, this information is limited.

Applicability

POCIS can be deployed in a wide range of aquatic environments including stagnant pools, rivers, springs, estuarine systems, and wastewater streams. However, there has been little research into the use of POCIS in strictly marine environments. Prior to deployment of a POCIS device, it is essential to select a study site that will maximize the effectiveness of the sampler. Selecting an area that is shaded will help prevent light sensitive chemicals from being degrading. The site should also allow the sampler to be submerged in the water without being buried in the sediment. It is ideal to place the sampler in moving water in order to increase sampling rates, however, areas with an extremely turbulent water flow should be avoided as to prevent damage to the POCIS device. Passive samplers are very vulnerable to vandalism and it is therefore important to secure the sampler in areas that are not easily visible and that are away from areas frequently used by people.

POCIS samplers can be deployed for a period of time ranging from weeks to months. The shortest deployment lengths are typically 7 days but average 2–3 months. It is important to have a long enough deployment period to allow for adequate detection of contaminants at ambient environmental concentrations. Often, the two different types of POCIS devices will be deployed together in order to provide the greatest understanding of contamination. It is also important to deploy enough POCIS devices to ensure a large enough sample of contaminant is recovered for chemical analysis. An estimate or the number of samplers needed at a given site can be determined by the following equation.

$$R_s \text{ x } t \text{ x } n \text{ x } C_c \text{ x } P_r \text{ x } E_t > MQL \text{ x } V_i$$

where

- C_c is the predicted environmental concentration of the contaminant

- t is the deployment time in days

- R_s is sampling rate in liters of water extracted by the passive sampler per day(L/day)

- P_r is the overall method recovery for the analyte (expressed as a factor of one; ::therefore 0.9 is used for 90 percent recovery),

- n is the number of passive samplers combined into a single sample,

- E_t is the fraction of the total sample extract which is injected into the ::instrument for quantification

- MQL is the method quantification limit

- V_i is the volume of standard injection (commonly 1 μL).

Relevant Contaminants

Any compound with a log Kow of less than or equal to 3 can concentrate in a POCIS sampler. Applicable classes of contaminants measured by POCIS are pharmaceuticals, household and industrial products, hormones, herbicides, and polar pesticides (Table 1). Currently, there are two POCIS configurations that are targeted for different classes of contaminants. A general POCIS design contains a sorbent that is used to collect pesticides, natural as well as synthetic hormones, and wastewater related chemicals. The pharmaceutical POCIS configuration contains a sorbent that is designed to specifically target classes of pharmaceuticals.

Applicable Contaminants that Concentrate in a POCIS Device. Not to be Considered a Complete List.

Chemical Class	Examples
Pharmaceuticals	acetaminophen, azithromycin, carbamazepine, ibuprofen, propranolol, sulfa drugs, tetracycline antibiotics
Household and industrial products	alkyl phenols, benzophenone, caffeine, DEET, fire retardants, indole, triclosan
Hormones	17β-estradiol, 17α-ethynlestradiol, estrone, estriol
Herbicides	atrazine, cyanazine, hydroxyatrazine, tertbutylazine
Polar pesticides	alachlor, chlorppyrifos, diainon, dichlorvos, diuron, isoproturon, metolachlor
Other	Urobilin

POCIS Processing

Before the POCIS is constructed, all the hardware as well as the sorbents and membrane must be thoroughly cleaned so that any potential interference is removed. During and after sampling the only cleaning necessary is the removal of any sediment that has adhered to the surface of the sampler. After assembly, and prior to deployment, the samplers are stored in frozen airtight containers to avoid any contamination. The samplers should be kept in airtight containers during transportation both to and from the sampling site so that airborne contaminants do not contaminate the sampler. It is ideal to keep the samplers cold while transporting them in order to preserve the integrity of the samples.

After the POCIS is retrieved from the field, the membrane is gently cleaned to reduce the possibility of any contamination to the sorbent. The sorbent is placed into a chromatography column so that the chemicals that samples can be recovered using an organic solvent. The solvent used is specifically chosen based on the type of sorbent and chemicals sampled. The sample can go through further processing such as cleanup or fractionation depending on the desired use of the sample.

Data Analysis

After the sample has been processed, the extract can be analysed using a variety of data analysis techniques. The chemical analysis and analytical instrumentation used depends on the goal of the study. Many analyses require multiple samples, although in some cases a single POCIS sample can be used for multiple analyses.

It is vital to use quality control (QC) procedures when using passive samplers. It is common practice for 10% to 50% of the total number of samples to be used for QC purposes. The number of QC samples depends on the study objectives. The QC samples are used to address issues such as sample contamination and analyte recovery. The types of QC samples commonly used include; reagent blanks, field blanks, matrix spikes, and procedural spikes.

A large number of studies have been performed in which POCIS data was combined with bioassays to measure biological endpoints. Testing POCIS extracts in biological assays is useful as a POCIS device samples over its entire deployment period, and biologically active compounds can be effectively monitored. It can also be argued that the use of POCIS is a more relevant from an ecotoxicological perspective as the use of a passive sampler mimics the uptake of compounds by organisms. Another strength in using bioassays to test environmental samples is that they can provide an integrative measure of the toxic potential of a group of chemical compounds, rather than a single contaminant.

Other Passive Samplers

There are many types of passive samplers used that specialize in absorbing different classes of aquatic contaminants found in the environment. Chemcatcher and SMPD are two types of passive samplers that are also commonly used. Monitoring programs use SMPDs to measure to hydrophobic organic contaminants. SPMDs are designed to mimic the bioconcentration of contaminants in fatty tissues (ITRC, 2006). Contaminants applicable to the of an SPMD include, but are not limited to, polychlorinated biphenyls (PCBs), polycyclic aromatic hydrocarbons (PAHs), organochlorine pesticides, dioxins, and furans.

The SPMD consist of a thin-walled, nonporous, polyethylene membrane tube that is filled with high molecular weight lipid. These tubes are approximately 90 cm long and wrap around the inside of a stainless steel deployment canister. SMPDs are efficient at absorbing pollutants with a log Kow of 4-8. This slightly overlaps with the range of contaminants absorbed by POCIS. Because of this, SMPDs and POCIS devices are often used together in monitoring studies to achieve a more representative understanding of contamination.

Future Development

The POCIS system is continually evaluated for the potential to sample a wide range of contaminants. Calibration data and analyte recovery methods are currently being generated by researchers around the world. Techniques to merge the POCIS device with bioassays are also under development. The POCIS sampler already serves as a versatile, economical, and robust tool for monitoring studies and observing trends in both space and time. However, sampling rates are not yet robust enough to supply reliable contaminant concentrations, particularly when regarding environmental quality standards. A limited number of sampling rates have been determined for chemicals and the determination of additional sampling rate data is necessary for the advancement of passive sampling technology.

Sorbent Tube

Sorbent tubes are the most widely used collection media for sampling hazardous gases and vapors in air, mostly as it relates to Industrial hygiene. They were developed by the US National Institute for Occupational Safety and Health (NIOSH) for air quality testing of workers. Sorbent Tubes are available from SKC Inc., 7Solutions BV, Uniphos Ltd., SKC Ltd, Zefon International, Sigma-Aldrich/Supelco and Markes International. SKC Inc. manufactured the first commercially available sorbent tubes.

Sorbent tubes are typically made of glass and contain various types of solid adsorbent material (sorbents). Commonly used sorbents include activated charcoal, silica gel, and organic porous polymers such as Tenax and Amberlite XAD resins. Solid sorbents are selected for sampling specific compounds in air because they:

1. Trap and retain the compound(s) of interest even in the presence of other compounds

2. Do not alter the compound(s) of interest

3. Allow collected compounds to be easily desorbed or extracted for analysis

Sorbent tubes are attached to air sampling pumps for sample collection. A pump with a calibrated flow rate in ml/min is normally placed on a worker's belt and it draws a known volume of air through the sorbent tube. Alternatively, pumps and sorbent tubes are placed in areas for fixed-point sampling. Chemicals are trapped onto the sorbent material throughout the sampling period.

Occasionally, when desorbing the air sample from the sorbent tube, a large portion of the analyte will fail to go into the solution. In these cases, the sorbent tubes will have to be adjusted for desorption efficiency (DE).

References

- T. Slocum, R. McMaster, F. Kessler, and H. Howard, Thematic Cartography and Geographic Visualization, 2nd edition, Pearson, 2005, ISBN 0-13-035123-7, p. 272.

- C. Close, The Early Years of the Ordnance Survey, 1926, republished by David and Charles, 1969, ISBN 0-7153-4477-3, pp. 141-144.

- Pavia, Donald L., Gary M. Lampman, George S. Kritz, Randall G. Engel (2006). Introduction to Organic Laboratory Techniques (4th Ed.). Thomson Brooks/Cole. pp. 797–817. ISBN 978-0-495-28069-9.

- Harris, Daniel C. (1999). "24. Gas Chromatography". Quantitative chemical analysis (Chapter) (Fifth ed.). W. H. Freeman and Company. pp. 675–712. ISBN 0-7167-2881-8.

- Grob, Robert L.; Barry, Eugene F. (2004). Modern Practice of Gas Chromatography (4th Ed.). John Wiley & Sons. ISBN 0-471-22983-0.

- Downard, Kevin (2004). "Mass Spectrometry - A Foundation Course". Royal Society of Chemistry. doi:10.1039/9781847551306. ISBN 978-0-85404-609-6.

- Eiceman, G.A. (2000). Gas Chromatography. In R.A. Meyers (Ed.), Encyclopedia of Analytical Chemistry: Applications, Theory, and Instrumentation, pp. 10627. Chichester: Wiley. ISBN 0-471-97670-9

- Tureček, František; McLafferty, Fred W. (1993). Interpretation of mass spectra. Sausalito: University Science Books. ISBN 0-935702-25-3.

- "Explaining road transport emissions - A non-technical guide, pages 27 to 37. — European Environment Agency". www.eea.europa.eu. Retrieved 2016-03-01.

- Johson, Dennis (2002-02-13). "ROVER - Real-time On-road Vehicle Emissions Reporter Dennis Johnson, US EPA" (PDF). Real-time On-road Vehicle Emissions Reporter Dennis Johnson, US EPA. US EPA. Retrieved 2016-03-01.

Environmental Monitoring in Air Pollution

Monitoring in air pollution can be done in various ways. Governments to communicate the present population of air in a particular area use air quality index. Some other ways of environmental monitoring are air quality index, upper-atmospheric models, indoor air quality, wind rose and atmospheric dispersion modelling. This section on environmental monitoring offers an insightful focus, keeping in mind the complex subject matter.

Air Quality Index

An air quality index (AQI) is a number used by government agencies to communicate to the public how polluted the air currently is or how polluted it is forecast to become. As the AQI increases, an increasingly large percentage of the population is likely to experience increasingly severe adverse health effects. Different countries have their own air quality indices, corresponding to different national air quality standards. Some of these are the Air Quality Health Index (Canada), the Air Pollution Index (Malaysia), and the Pollutant Standards Index (Singapore).

Smog builds up under an inversion in Almaty, Kazakhstan resulting in a high AQI

Definition and Usage

Computation of the AQI requires an air pollutant concentration over a specified averaging period, obtained from an air monitor or model. Taken together, concentration and time represent the dose of the air pollutant. Health effects corresponding to a given dose are established by epidemiological research. Air pollutants vary in potency, and the function used to convert from air pollutant concentration to AQI varies by pollutant. Air quality index values are typically grouped into ranges. Each range is assigned a descriptor, a color code, and a standardized public health advisory.

An air quality measurement station in Edinburgh, Scotland

The AQI can increase due to an increase of air emissions (for example, during rush hour traffic or when there is an upwind forest fire) or from a lack of dilution of air pollutants. Stagnant air, often caused by an anticyclone, temperature inversion, or low wind speeds lets air pollution remain in a local area, leading to high concentrations of pollutants, chemical reactions between air contaminants and hazy conditions.

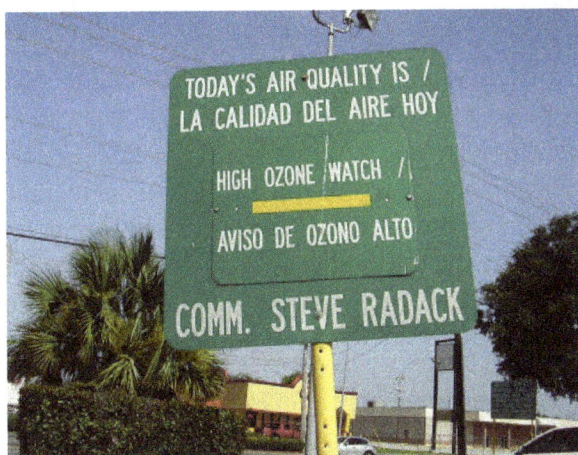

Signboard in Gulfton, Houston indicating an ozone watch

On a day when the AQI is predicted to be elevated due to fine particle pollution, an agency or public health organization might:

- advise sensitive groups, such as the elderly, children, and those with respiratory or cardiovascular problems to avoid outdoor exertion.

- declare an "action day" to encourage voluntary measures to reduce air emissions, such as using public transportation.

- recommend the use of masks to keep fine particles from entering the lungs

During a period of very poor air quality, such as an air pollution episode, when the AQI indicates that acute exposure may cause significant harm to the public health, agencies may invoke emergency plans that allow them to order major emitters (such as coal burning industries) to curtail emissions until the hazardous conditions abate.

Most air contaminants do not have an associated AQI. Many countries monitor ground-level ozone, particulates, sulfur dioxide, carbon monoxide and nitrogen dioxide, and calculate air quality indices for these pollutants.

The definition of the AQI in a particular nation reflects the discourse surrounding the development of national air quality standards in that nation. A website allowing government agencies anywhere in the world to submit their real-time air monitoring data for display using a common definition of the air quality index has recently become available.

Indices by Location

Canada

Air quality in Canada has been reported for many years with provincial Air Quality Indices (AQIs). Significantly, AQI values reflect air quality management objectives, which are based on the lowest achievable emissions rate, and not exclusively concern for human health. The Air Quality Health Index or (AQHI) is a scale designed to help understand the impact of air quality on health. It is a health protection tool used to make decisions to reduce short-term exposure to air pollution by adjusting activity levels during increased levels of air pollution. The Air Quality Health Index also provides advice on how to improve air quality by proposing behavioural change to reduce the environmental footprint. This index pays particular attention to people who are sensitive to air pollution. It provides them with advice on how to protect their health during air quality levels associated with low, moderate, high and very high health risks.

The Air Quality Health Index provides a number from 1 to 10+ to indicate the level of health risk associated with local air quality. On occasion, when the amount of air pollution is abnormally high, the number may exceed 10. The AQHI provides a local air quality current value as well as a local air quality maximums forecast for today, tonight, and tomorrow, and provides associated health advice.

| 1 | 2 | 3 | 4 | 5 | 6 | 7 | 8 | 9 | 10 | + |

Risk:	Low (1–3)	Moderate (4–6)	High (7–10)	Very high (above 10)

Health Risk	Air Quality Health Index	Health Messages	
		At Risk population	***General Population**
Low	1–3	Enjoy your usual outdoor activities.	Ideal air quality for outdoor activities
Moderate	4–6	Consider reducing or rescheduling strenuous activities outdoors if you are experiencing symptoms.	No need to modify your usual outdoor activities unless you experience symptoms such as coughing and throat irritation.
High	7–10	Reduce or reschedule strenuous activities outdoors. Children and the elderly should also take it easy.	Consider reducing or rescheduling strenuous activities outdoors if you experience symptoms such as coughing and throat irritation.
Very high	Above 10	Avoid strenuous activities outdoors. Children and the elderly should also avoid outdoor physical exertion.	Reduce or reschedule strenuous activities outdoors, especially if you experience symptoms such as coughing and throat irritation.

Hong Kong

On the 30th December 2013 Hong Kong replaced the Air Pollution Index with a new index called the *Air Quality Health Index*. This index is on a scale of 1 to 10+ and considers four air pollutants: ozone; nitrogen dioxide; sulphur dioxide and particulate matter (including PM10 and PM2.5). For any given hour the AQHI is calculated from the sum of the percentage excess risk of daily hospital admissions attributable to the 3-hour moving average concentrations of these four pollutants. The AQHIs are grouped into five AQHI health risk categories with health advice provided:

Health risk category	AQHI
Low	1
	2
	3
Medium	4
	5
	6
High	7
Very High	8
	9
	10
Serious	10+

Each of the health risk categories has advice with it. At the *low* and *moderate* levels the public are advised that they can continue normal activities. For the *high* category, children, the elderly and people with heart or respiratory illnesses are advising to reduce outdoor physical exertion. Above this (*very high* or *serious*) the general public are also advised to reduce or avoid outdoor physical exertion.

Mainland China

China's Ministry of Environmental Protection (MEP) is responsible for measuring the level of air pollution in China. As of 1 January 2013, MEP monitors daily pollution level in 163 of its major cities. The API level is based on the level of 6 atmospheric pollutants, namely sulfur dioxide (SO_2), nitrogen dioxide (NO_2), suspended particulates smaller than 10 μm in aerodynamic diameter (PM_{10}), suspended particulates smaller than 2.5 μm in aerodynamic diameter ($PM_{2.5}$), carbon monoxide (CO), and ozone (O_3) measured at the monitoring stations throughout each city.

AQI Mechanics

An individual score (IAQI) is assigned to the level of each pollutant and the final AQI is the highest of those 6 scores. The pollutants can be measured quite differently. $PM_{2.5}$, PM_{10} concentration are measured as average per 24h. SO_2, NO_2, O_3, CO are measured as average per hour. The final API value is calculated per hour according to a formula published by the MEP.

The scale for each pollutant is non-linear, as is the final AQI score. Thus an AQI of 100 does not mean twice the pollution of AQI at 50, nor does it mean twice as harmful. While an AQI of 50 from day 1 to 182 and AQI of 100 from day 183 to 365 does provide an annual average of 75, it does *not* mean the pollution is acceptable even if the benchmark of 100 is deemed safe. This is because the benchmark is a 24-hour target. The annual average must match against the annual target. It is entirely possible to have safe air every day of the year but still fail the annual pollution benchmark.

AQI and Health Implications (HJ 663-2012)

AQI	Air Pollution Level	Health Implications
0–50	Excellent	No health implications.
51–100	Good	Few hypersensitive individuals should reduce outdoor exercise.
101–150	Lightly Polluted	Slight irritations may occur, individuals with breathing or heart problems should reduce outdoor exercise.
151–200	Moderately Polluted	Slight irritations may occur, individuals with breathing or heart problems should reduce outdoor exercise.
201–300	Heavily Polluted	Healthy people will be noticeably affected. People with breathing or heart problems will experience reduced endurance in activities. These individuals and elders should remain indoors and restrict activities.

300+	Severely Polluted	Healthy people will experience reduced endurance in activities. There may be strong irritations and symptoms and may trigger other illnesses. Elders and the sick should remain indoors and avoid exercise. Healthy individuals should avoid outdoor activities.

India

The Minister for Environment, Forests & Climate Change Shri Prakash Javadekar launched The National Air Quality Index (AQI) in New Delhi on 17 September 2014 under the Swachh Bharat Abhiyan. It is outlined as 'One Number- One Colour-One Description' for the common man to judge the air quality within his vicinity. The index constitutes part of the Government's mission to introduce the culture of cleanliness. Institutional and infrastructural measures are being undertaken in order to ensure that the mandate of cleanliness is fulfilled across the country and the Ministry of Environment, Forests & Climate Change proposed to discuss the issues concerned regarding quality of air with the Ministry of Human Resource Development in order to include this issue as part of the sensitisation programme in the course curriculum.

While the earlier measuring index was limited to three indicators, the current measurement index had been made quite comprehensive by the addition of five additional parameters. Under the current measurement of air quality there are 8 parameters. The initiatives undertaken by the Ministry recently aimed at balancing environment and conservation and development as air pollution has been a matter of environmental and health concerns, particularly in urban areas.

The Central Pollution Control Board along with State Pollution Control Boards has been operating National Air Monitoring Program (NAMP) covering 240 cities of the country having more than 342 monitoring stations. In addition, continuous monitoring systems that provide data on near real-time basis are also installed in a few cities. They provide information on air quality in public domain in simple linguistic terms that is easily understood by a common person. Air Quality Index (AQI) is one such tool for effective dissemination of air quality information to people. As such an Expert Group comprising medical professionals, air quality experts, academia, advocacy groups, and SPCBs was constituted and a technical study was awarded to IIT Kanpur. IIT Kanpur and the Expert Group recommended an AQI scheme in 2014.

There are six AQI categories, namely Good, Satisfactory, Moderately polluted, Poor, Very Poor, and Severe. The proposed AQI will consider eight pollutants (PM_{10}, $PM_{2.5}$, NO_2, SO_2, CO, O_3, NH_3, and Pb) for which short-term (up to 24-hourly averaging period) National Ambient Air Quality Standards are prescribed. Based on the measured ambient concentrations, corresponding standards and likely health impact, a sub-index is calculated for each of these pollutants. The worst sub-index reflects overall AQI. Associated likely health impacts for different AQI categories and pollutants have been also been suggested, with primary inputs from the medical expert members of the group.

The AQI values and corresponding ambient concentrations (health breakpoints) as well as associated likely health impacts for the identified eight pollutants are as follows:

AQI Category, Pollutants and Health Breakpoints								
AQI Category (Range)	PM_{10} (24hr)	$PM_{2.5}$ (24hr)	NO_2 (24hr)	O_3 (8hr)	CO (8hr)	SO_2 (24hr)	NH_3 (24hr)	Pb (24hr)
Good (0-50)	0-50	0-30	0-40	0-50	0-1.0	0-40	0-200	0-0.5
Satisfactory (51-100)	51-100	31-60	41-80	51-100	1.1-2.0	41-80	201-400	0.5-1.0
Moderately polluted (101-200)	101-250	61-90	81-180	101-168	2.1-10	81-380	401-800	1.1-2.0
Poor (201-300)	251-350	91-120	181-280	169-208	10-17	381-800	801-1200	2.1-3.0
Very poor (301-400)	351-430	121-250	281-400	209-748	17-34	801-1600	1200-1800	3.1-3.5
Severe (401-500)	430+	250+	400+	748+	34+	1600+	1800+	3.5+

AQI	Associated Health Impacts
Good (0-50)	Minimal impact
Satisfactory (51-100)	May cause minor breathing discomfort to sensitive people.
Moderately polluted (101–200)	May cause breathing discomfort to people with lung disease such as asthma, and discomfort to people with heart disease, children and older adults.
Poor (201-300)	May cause breathing discomfort to people on prolonged exposure, and discomfort to people with heart disease.
Very poor (301-400)	May cause respiratory illness to the people on prolonged exposure. Effect may be more pronounced in people with lung and heart diseases.
Severe (401-500)	May cause respiratory impact even on healthy people, and serious health impacts on people with lung/heart disease. The health impacts may be experienced even during light physical activity.

Mexico

The air quality in Mexico City is reported in IMECAs. The IMECA is calculated using the measurements of average times of the chemicals ozone (O_3), sulphur dioxide (SO_2), nitrogen dioxide (NO_2), carbon monoxide (CO), particles smaller than 2.5 micrometers ($PM_{2.5}$), and particles smaller than 10 micrometers (PM_{10}).

Singapore

Singapore uses the Pollutant Standards Index to report on its air quality, with details of the calculation similar but not identical to that used in Malaysia and Hong Kong The PSI chart below is grouped by index values and descriptors, according to the National Environment Agency.

PSI	Descriptor	General Health Effects
0–50		None
51–100	Moderate	Few or none for the general population
101–200	Unhealthy	Mild aggravation of symptoms among susceptible persons i.e. those with underlying conditions such as chronic heart or lung ailments; transient symptoms of irritation e.g. eye irritation, sneezing or coughing in some of the healthy population.
201–300	Very Unhealthy	Moderate aggravation of symptoms and decreased tolerance in persons with heart or lung disease; more widespread symptoms of transient irritation in the healthy population.
301–400	Hazardous	Early onset of certain diseases in addition to significant aggravation of symptoms in susceptible persons; and decreased exercise tolerance in healthy persons.
Above 400	Hazardous	PSI levels above 400 may be life-threatening to ill and elderly persons. Healthy people may experience adverse symptoms that affect normal activity.

South Korea

The Ministry of Environment of South Korea uses the Comprehensive Air-quality Index (CAI) to describe the ambient air quality based on the health risks of air pollution. The index aims to help the public easily understand the air quality and protect people's health. The CAI is on a scale from 0 to 500, which is divided into six categories. The higher the CAI value, the greater the level of air pollution. Of values of the five air pollutants, the highest is the CAI value. The index also has associated health effects and a colour representation of the categories as shown below.

CAI	Description	Health Implications
0–50	Good	A level that will not impact patients suffering from diseases related to air pollution.
51–100	Moderate	A level that may have a meager impact on patients in case of chronic exposure.
101–150	Unhealthy for sensitive groups	A level that may have harmful impacts on patients and members of sensitive groups.
151–250	Unhealthy	A level that may have harmful impacts on patients and members of sensitive groups (children, aged or weak people), and also cause the general public unpleasant feelings.
251–500	Very unhealthy	A level that may have a serious impact on patients and members of sensitive groups in case of acute exposure.

The N Seoul Tower on Namsan Mountain in central Seoul, South Korea, is illuminated in blue, from sunset to 23:00 and 22:00 in winter, on days where the air quality in Seoul is 45 or less. During the spring of 2012, the Tower was lit up for 52 days, which is four days more than in 2011.

United Kingdom

The most commonly used air quality index in the UK is the *Daily Air Quality Index* recommended by the Committee on Medical Effects of Air Pollutants (COMEAP). This index has ten points, which are further grouped into 4 bands: low, moderate, high and very high. Each of the bands comes with advice for at-risk groups and the general population.

Air pollution banding	Value	Health messages for At-risk individuals	Health messages for General population
Low	1–3	Enjoy your usual outdoor activities.	Enjoy your usual outdoor activities.
Moderate	4–6	Adults and children with lung problems, and adults with heart problems, who experience symptoms, should consider reducing strenuous physical activity, particularly outdoors.	Enjoy your usual outdoor activities.
High	7–9	Adults and children with lung problems, and adults with heart problems, should reduce strenuous physical exertion, particularly outdoors, and particularly if they experience symptoms. People with asthma may find they need to use their reliever inhaler more often. Older people should also reduce physical exertion.	Anyone experiencing discomfort such as sore eyes, cough or sore throat should consider reducing activity, particularly outdoors.
Very High	10	Adults and children with lung problems, adults with heart problems, and older people, should avoid strenuous physical activity. People with asthma may find they need to use their reliever inhaler more often.	Reduce physical exertion, particularly outdoors, especially if you experience symptoms such as cough or sore throat.

The index is based on the concentrations of 5 pollutants. The index is calculated from the concentrations of the following pollutants: Ozone, Nitrogen Dioxide, Sulphur Dioxide, PM2.5 (particles with an aerodynamic diameter less than 2.5 μm) and PM10. The breakpoints between index values are defined for each pollutant separately and the overall index is defined as the maximum value of the index. Different averaging periods are used for different pollutants.

Index	Ozone, Running 8 hourly mean (μg/m³)	Nitrogen Dioxide, Hourly mean (μg/m³)	Sulphur Dioxide, 15 minute mean (μg/m³)	PM2.5 Particles, 24 hour mean (μg/m³)	PM10 Particles, 24 hour mean (μg/m³)
1	0-33	0-67	0-88	0-11	0-16
2	34-66	68-134	89-177	12-23	17-33
3	67-100	135-200	178-266	24-35	34-50
4	101-120	201-267	267-354	36-41	51-58
5	121-140	268-334	355-443	42-47	59-66
6	141-160	335-400	444-532	48-53	67-75
7	161-187	401-467	533-710	54-58	76-83

8	188-213	468-534	711-887	59-64	84-91
9	214-240	535-600	888-1064	65-70	92-100
10	≥ 241	≥ 601	≥ 1065	≥ 71	≥ 101

Europe

To present the air quality situation in European cities in a comparable and easily understandable way, all detailed measurements are transformed into a single relative figure: the Common Air Quality Index (or CAQI) Three different indices have been developed by Citeair to enable the comparison of three different time scale:.

- An hourly index, which describes the air quality today, based on hourly values and updated every hours,

- A daily index, which stands for the general air quality situation of yesterday, based on daily values and updated once a day,

- An annual index, which represents the city's general air quality conditions throughout the year and compare to European air quality norms. This index is based on the pollutants year average compare to annual limit values, and updated once a year.

However, the proposed indices and the supporting common web site www.airqualitynow.eu are designed to give a dynamic picture of the air quality situation in each city but not for compliance checking.

The Hourly and Daily Common Indices

These indices have 5 levels using a scale from 0 (very low) to > 100 (very high), it is a relative measure of the amount of air pollution. They are based on 3 pollutants of major concern in Europe: PM10, NO2, O3 and will be able to take into account to 3 additional pollutants (CO, PM2.5 and SO2) where data are also available.

The calculation of the index is based on a review of a number of existing air quality indices, and it reflects EU alert threshold levels or daily limit values as much as possible. In order to make cities more comparable, independent of the nature of their monitoring network two situations are defined:

- Background, representing the general situation of the given agglomeration (based on urban background monitoring sites),

- Roadside, being representative of city streets with a lot of traffic, (based on roadside monitoring stations)

The indices values are updated hourly (for those cities that supply hourly data) and yesterdays daily indices are presented.

Common air quality index legend:

Pollution	Index Value
Very low	0/25
Low	25/50
Medium	50/75
High	75/100
Very high	>100

The Common Annual Air Quality Index

The common annual air quality index provides a general overview of the air quality situation in a given city all the year through and regarding to the European norms.

It is also calculated both for background and traffic conditions but its principle of calculation is different from the hourly and daily indices. It is presented as a distance to a target index, this target being derived from the EU directives (annual air quality standards and objectives):

- If the index is higher than 1: for one or more pollutants the limit values are not met.

- If the index is below 1: on average the limit values are met.

The annual index is aimed at better taking into account long term exposure to air pollution based on distance to the target set by the EU annual norms, those norms being linked most of the time to recommendations and health protection set up by World Health Organisation.

United States

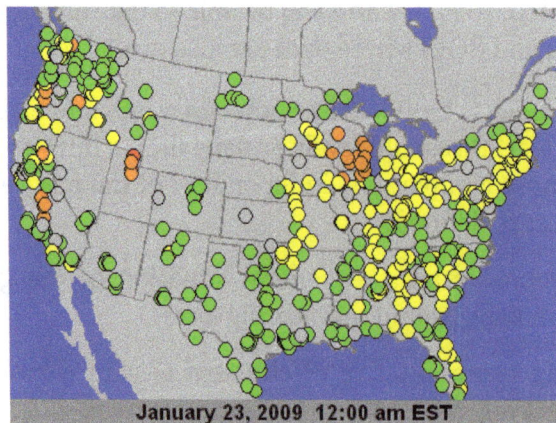

January 23, 2009 12:00 am EST

PM$_{2.5}$ 24-Hour AQI Loop, Courtesy US EPA

The United States Environmental Protection Agency (EPA) has developed an Air Quality Index that is used to report air quality. This AQI is divided into six categories indicating increasing levels of health concern. An AQI value over 300 represents hazardous air quality and below 50 the air quality is good.

Air Quality Index (AQI) Values	Levels of Health Concern	Colors
0 to 50	Good	Green
51 to 100	Moderate	Yellow
101 to 150	Unhealthy for Sensitive Groups	Orange
151 to 200	Unhealthy	Red
201 to 300	Very Unhealthy	Purple
301 to 500	Hazardous	Maroon

The AQI is based on the five "criteria" pollutants regulated under the Clean Air Act: ground-level ozone, particulate matter, carbon monoxide, sulfur dioxide, and nitrogen dioxide. The EPA has established National Ambient Air Quality Standards (NAAQS) for each of these pollutants in order to protect public health. An AQI value of 100 generally corresponds to the level of the NAAQS for the pollutant. The Clean Air Act (USA) (1990) requires EPA to review its National Ambient Air Quality Standards every five years to reflect evolving health effects information. The Air Quality Index is adjusted periodically to reflect these changes.

Computing the AQI

The air quality index is a piecewise linear function of the pollutant concentration. At the boundary between AQI categories, there is a discontinuous jump of one AQI unit. To convert from concentration to AQI this equation is used:

$$I = \frac{I_{high} - I_{low}}{C_{high} - C_{low}}(C - C_{low}) + I_{low}$$

where:

I = the (Air Quality) index,

C = the pollutant concentration,

C_{low} = the concentration breakpoint that is $\leq C$,

C_{high} = the concentration breakpoint that is $\geq C$,

I_{low} = the index breakpoint corresponding to C_{low},,

I_{high} = the index breakpoint corresponding to C_{high}.

EPA's table of breakpoints is:

Suppose a monitor records a 24-hour average fine particle ($PM_{2.5}$) concentration of 12.0 micrograms per cubic meter. The equation above results in an AQI of:

$$\frac{50-0}{12.0-0}(12.0-0)+0=50,$$

corresponding to air quality in the "Good" range. To convert an air pollutant concentration to an AQI, EPA has developed a calculator.

If multiple pollutants are measured at a monitoring site, then the largest or "dominant" AQI value is reported for the location. The ozone AQI between 100 and 300 is computed by selecting the larger of the AQI calculated with a 1-hour ozone value and the AQI computed with the 8-hour ozone value.

8-hour ozone averages do not define AQI values greater than 300; AQI values of 301 or greater are calculated with 1-hour ozone concentrations. 1-hour SO_2 values do not define higher AQI values greater than 200. AQI values of 201 or greater are calculated with 24-hour SO_2 concentrations.

Real time monitoring data from continuous monitors are typically available as 1-hour averages. However, computation of the AQI for some pollutants requires averaging over multiple hours of data. (For example, calculation of the ozone AQI requires computation of an 8-hour average and computation of the $PM_{2.5}$ or PM_{10} AQI requires a 24-hour average.) To accurately reflect the current air quality, the multi-hour average used for the AQI computation should be centered on the current time, but as concentrations of future hours are unknown and are difficult to estimate accurately, EPA uses surrogate concentrations to estimate these multi-hour averages. For reporting the $PM_{2.5}$, PM_{10} and ozone air quality indices, this surrogate concentration is called the NowCast. The Nowcast is a particular type of weighted average that provides more weight to the most recent air quality data when air pollution levels are changing.

Public Availability of the AQI

Real time monitoring data and forecasts of air quality that are color-coded in terms of the air quality index are available from EPA's AirNow web site. Historical air monitoring data including AQI charts and maps are available at EPA's AirData website.

History of the AQI

The AQI made its debut in 1968, when the National Air Pollution Control Administration undertook an initiative to develop an air quality index and to apply the methodology to Metropolitan Statistical Areas. The impetus was to draw public attention to the issue of air pollution and indirectly push responsible local public officials to take action to control sources of pollution and enhance air quality within their jurisdictions.

Jack Fensterstock, the head of the National Inventory of Air Pollution Emissions and Control Branch, was tasked to lead the development of the methodology and to compile the air quality and emissions data necessary to test and calibrate resultant indices.

The initial iteration of the air quality index used standardized ambient pollutant concentrations to yield individual pollutant indices. These indices were then weighted and summed to form a single total air quality index. The overall methodology could use concentrations that are taken from ambient monitoring data or are predicted by means of a diffusion model. The concentrations were then converted into a standard statistical distribution with a preset mean and standard deviation. The resultant individual pollutant indices are assumed to be equally weighted, although values other than unity can be used. Likewise, the index can incorporate any number of pollutants although it was only used to combine SOx, CO, and TSP because of a lack of available data for other pollutants.

While the methodology was designed to be robust, the practical application for all metropolitan areas proved to be inconsistent due to the paucity of ambient air quality monitoring data, lack of agreement on weighting factors, and non-uniformity of air quality standards across geographical and political boundaries. Despite these issues, the publication of lists ranking metropolitan areas achieved the public policy objectives and led to the future development of improved indices and their routine application.

Upper-atmospheric Models

Most climate models simulate a region of the Earth's atmosphere from the surface to the stratopause. There also exist numerical models which simulate the wind, temperature and composition of the Earth's tenuous upper atmosphere, from the mesosphere to the exosphere, including the ionosphere. This region is affected strongly by the 11 year Solar cycle through variations in solar UV/EUV/Xray radiation and solar wind leading to high latitude particle precipitation and aurora. It has been proposed that these phenomena may have an effect on the lower atmosphere, and should therefore be included in simulations of climate change. For this reason there has been a drive in recent years to create "whole atmosphere" models to investigate whether or not this is the case.

A jet stream perturbation model is employed by Weather Logistics UK, which simulates the diversion of the air streams in the upper atmosphere. North Atlantic air flow modelling is simulated by combining a monthly jet stream climatology input calculated at 20 to 30°W, with different blocking high patterns. The jet stream input is generated by thermal wind balance calculations at 316mbars (6 to 9 km aloft) in the mid-latitude range from 40 to 60°N. Long term blocking patterns are determined by the weather forecaster, who identifies the likely position and strength of North Atlantic Highs from synoptic charts, the North Atlantic Oscillation (NAO) and El Niño-Southern Oscillation (ENSO) patterns. The model is based on the knowledge that low pressure systems at the surface are steered by the fast ribbons (jet streams) of air in the upper atmosphere. The jet stream - blocking interaction model simulation examines the sea surface temperature field using data from NOAA tracked along the ocean on a path to the British Isles. The principal theory suggests that long term weather patterns act on longer time scales, so large blocking patterns are thought to appear in a similar locations repeatedly over several months. With a good knowledge of blocking high patterns, the model performs with an impressive accuracy that is useful to the end user.

The modelling undertaken at Weather Logistics UK produces regional-seasonal predictions that are probabilistic in nature. Two different blocking sizes are used for the modelling, located at two different locations. The four possible blocking diversions are then ranked in an order, to be combined by logistic regression and generate the appropriate likelihoods of weather events on seasonal time-scales. The raw output consists of 22 different weather conditions for each season that are compared to the average atmospheric conditions. A global warming bias and 1961–1990 climatology of regional British Isles temperatures are added to the anomaly value to produce a final temperature prediction. The seasonal weather forecasts at Weather Logistics UK include several additional weather components (derivatives) including: precipitation anomalies, storm tracks, air flow trajectories, heating degree days for household utility bills, cooling degree days, heat wave and the snow day odds.

According to a report in New Scientist many researchers are in consensus that Rossby waves are acting against the jet stream's usual pattern and holding it in place. Upper atmospheric studies using National Oceanic and Atmospheric Administration (NOAA) data indicates that during July 2010 these upper air stream patterns were most frequently observed in the Northern Hemisphere. Examination of the climatology data over the same period of time indicates that these wild planetary wave meanderings are not a normal aspect of our regional climate patterns. Meanwhile, ongoing research studies at the University of Reading show that unusual patterns in the polar jet stream are more common during a period of low activity in the solar cycle when the observed sun spot activity and their associated solar flares are at their minimum. The link between low solar activity and enhanced blocking patterns is associated with an increase in the prevalence of cold weather patterns during the European Winter. Another possible explanation for the observed increase in blocking patterns is natural variability, through the chaotic character of the large-scale Ocean currents that flow across the surface of the tropical Pacific.

Indoor Air Quality

Indoor air quality (IAQ) is a term which refers to the air quality within and around buildings and structures, especially as it relates to the health and comfort of building occupants. IAQ can be affected by gases (including carbon monoxide, radon, volatile organic compounds), particulates, microbial contaminants (mold, bacteria), or any mass or energy stressor that can induce adverse health conditions. Source control, filtration and the use of ventilation to dilute contaminants are the primary methods for improving indoor air quality in most buildings. Residential units can further improve indoor air quality by routine cleaning of carpets and area rugs.

Determination of IAQ involves the collection of air samples, monitoring human exposure to pollutants, collection of samples on building surfaces, and computer modelling of air flow inside buildings.

IAQ is part of indoor environmental quality (IEQ), which includes IAQ as well as other physical and psychological aspects of life indoors (e.g., lighting, visual quality, acoustics, and thermal comfort).

Indoor air pollution in developing nations is a major health hazard. A major source of indoor air pollution in developing countries is the burning of biomass (e.g. wood, charcoal, dung, or crop residue) for heating and cooking. The resulting exposure to high levels of particulate matter resulted in between 1.5 million and 2 million deaths in 2000.

Common Pollutants

Second-hand Smoke

Second-hand smoke is tobacco smoke which affects other people other than the 'active' smoker. Second-hand tobacco smoke includes both a gaseous and a particulate phase, with particular hazards arising from levels of carbon monoxide (as indicated below) and very small particulates (at PM2.5 size) which get past the lung's natural defenses. [original research?] The only certain method to improve indoor air quality as regards second-hand smoke is the implementation of comprehensive smoke-free laws.

Radon

Radon is an invisible, radioactive atomic gas that results from the radioactive decay of radium, which may be found in rock formations beneath buildings or in certain building materials themselves. Radon is probably the most pervasive serious hazard for indoor air in the United States and Europe, probably responsible for tens of thousands of deaths from lung cancer each year. There are relatively simple test kits for do-it-yourself radon gas testing, but if a home is for sale the testing must be done by licensed person in some U.S. states. Radon gas enters buildings as a soil gas and is a heavy gas and thus will tend to accumulate at the lowest level. Radon may also be introduced into a building through drinking water particularly from bathroom showers. Building materials can be a rare source of radon, but little testing is carried out for stone, rock or tile products brought into building sites; radon accumulation is greatest for well insulated homes. The half life for radon is 3.8 days, indicating that once the source is removed, the hazard will be greatly reduced within a few weeks. Radon mitigation methods include sealing concrete slab floors, basement foundations, water drainage systems, or by increasing ventilation. They are usually cost effective and can greatly reduce or even eliminate the contamination and the associated health risks.

Molds and Other Allergens

These biological chemicals can arise from a host of means, but there are two common classes: (a) moisture induced growth of mold colonies and (b) natural substances released into the air such as animal dander and plant pollen. Mold is always associated with moisture, and its growth can be inhibited by keeping humidity levels below 50%. Moisture buildup inside buildings may arise from water penetrating compromised areas of the building envelope or skin, from plumbing leaks, from condensation due to improper ventilation, or from ground moisture penetrating a building part. In areas where cellulosic materials (paper and wood, including drywall) become moist and fail to dry within 48 hours, mold mildew can propagate and release allergenic spores into the air.

In many cases, if materials have failed to dry out several days after the suspected water event, mold growth is suspected within wall cavities even if it is not immediately visible. Through a mold investigation, which may include destructive inspection, one should be able to determine the presence or absence of mold. In a situation where there is visible mold and the indoor air quality may have been compromised, mold remediation may be needed. Mold testing and inspections should be carried out by an independent investigator to avoid any conflict of interest and to insure accurate results; free mold testing offered by remediation companies is not recommended.

There are some varieties of mold that contain toxic compounds (mycotoxins). However, exposure to hazardous levels of mycotoxin via inhalation is not possible in most cases, as toxins are produced by the fungal body and are not at significant levels in the released spores. The primary hazard of mold growth, as it relates to indoor air quality, comes from the allergenic properties of the spore cell wall. More serious than most allergenic properties is the ability of mold to trigger episodes in persons that already have asthma, a serious respiratory disease.

Carbon Monoxide

One of the most acutely toxic indoor air contaminants is carbon monoxide (CO), a colourless, odourless gas that is a byproduct of incomplete combustion of fossil fuels.

Common sources of carbon monoxide are tobacco smoke, space heaters using fossil fuels, defective central heating furnaces and automobile exhaust. By depriving the brain of oxygen, high levels of carbon monoxide can lead to nausea, unconsciousness and death. According to the American Conference of Governmental Industrial Hygienists (ACGIH), the time-weighted average (TWA) limit for carbon monoxide (630-08-0) is 25 ppm.

Indoor levels of CO are systematically improving due to increasing implementation of smoke-free laws.

Volatile Organic Compounds

Volatile organic compounds (VOCs) are emitted as gases from certain solids or liquids. VOCs include a variety of chemicals, some of which may have short- and long-term adverse health effects. Concentrations of many VOCs are consistently higher indoors (up to ten times higher) than outdoors. VOCs are emitted by a wide array of products numbering in the thousands. Examples include: paints and lacquers, paint strippers, cleaning supplies, pesticides, building materials and furnishings, office equipment such as copiers and printers, correction fluids and carbonless copy paper, graphics and craft materials including glues and adhesives, permanent markers, and photographic solutions.

Chlorinated drinking water releases chloroform when hot water is used in the home. Benzene is emitted from fuel stored in attached garages. Overheated cooking oils emit acrolein and formaldehyde. A meta-analysis of 77 surveys of VOCs in homes in the US found the top ten riskiest indoor air VOCs were acrolein, formaldehyde, benzene, hexachlorobutadiene, acetaldehyde, 1,3-butadiene, benzyl chloride, 1,4-dichlorobenzene, carbon tetrachloride, acrylonitrile, and vinyl chloride. These compounds exceeded health standards in most homes.

Organic chemicals are widely used as ingredients in household products. Paints, varnishes, and wax all contain organic solvents, as do many cleaning, disinfecting, cosmetic, degreasing, and hobby products. Fuels are made up of organic chemicals. All of these products can release organic compounds during usage, and, to some degree, when they are stored. Testing emissions from building materials used indoors has become increasingly common for floor coverings, paints, and many other important indoor building materials and finishes.

Several initiatives envisage to reduce indoor air contamination by limiting VOC emissions from products. There are regulations in France and in Germany, and numerous voluntary ecolabels and rating systems containing low VOC emissions criteria such as EMICODE, M1, Blue Angel and Indoor Air Comfort in Europe, as well as California Standard CDPH Section 01350 and several others in the USA. These initiatives changed the marketplace where an increasing number of low-emitting products has become available during the last decades.

At least 18 Microbial VOCs (MVOCs) have been characterised including 1-octen-3-ol, 3-methylfuran, 2-pentanol, 2-hexanone, 2-heptanone, 3-octanone, 3-octanol, 2-octen-1-ol, 1-octene, 2-pentanone, 2-nonanone, borneol, geosmin, 1-butanol, 3-methyl-1-butanol, 3-methyl-2-butanol, and thujopsene. The first of these compounds is called mushroom alcohol. The last four are products of Stachybotrys chartarum, which has been linked with sick building syndrome.

Legionella

Legionellosis or Legionnaire's Disease is caused by a waterborne bacterium Legionella that grows best in slow-moving or still, warm water. The primary route of exposure is through the creation of an aerosol effect, most commonly from evaporative cooling towers or showerheads. A common source of Legionella in commercial buildings is from poorly placed or maintained evaporative cooling towers, which often release water in an aerosol which may enter nearby ventilation intakes. Outbreaks in medical facilities and nursing homes, where patients are immuno-suppressed and immuno-weak, are the most commonly reported cases of Legionellosis. More than one case has involved outdoor fountains in public attractions. The presence of Legionella in commercial building water supplies is highly under-reported, as healthy people require heavy exposure to acquire infection.

Legionella testing typically involves collecting water samples and surface swabs from evaporative cooling basins, shower heads, faucets/taps, and other locations where warm water collects. The samples are then cultured and colony forming units (cfu) of Legionella are quantified as cfu/Liter.

Legionella is a parasite of protozoans such as amoeba, and thus requires conditions suitable for both organisms. The bacterium forms a biofilm which is resistant to chemical and antimicrobial treatments, including chlorine. Remediation for Legionella outbreaks in commercial buildings vary, but often include very hot water flushes (160 °F; 70 °C), sterilisation of standing water in evaporative cooling basins, replacement of shower heads, and in some cases flushes of heavy metal salts. Preventative measures include adjusting normal hot water levels to allow for 120 °F at the tap, evaluating facility design layout, removing faucet aerators, and periodic testing in suspect areas.

Other Bacteria

There are many bacteria of health significance found in indoor air and on indoor surfaces. The role of microbes in the indoor environment is increasingly studied using modern gene-based analysis of environmental samples. Currently efforts are under way to link microbial ecologists and indoor air scientists to forge new methods for analysis and to better interpret the results.

Bacteria (26 2 27) Airborne Microbes

"There are approximately ten times as many bacterial cells in the human flora as there are human cells in the body, with large numbers of bacteria on the skin and as gut flora." A large fraction of the bacteria found in indoor air and dust are shed from humans. Among the most important bacteria known to occur in indoor air are Mycobacterium tuberculosis, Staphylococcus aureus, Streptococcus pneumoniae.

Asbestos Fibers

Many common building materials used before 1975 contain asbestos, such as some floor tiles, ceiling tiles, shingles, fireproofing, heating systems, pipe wrap, taping muds, mastics, and other insulation materials. Normally, significant releases of asbestos fiber do not occur unless the building materials are disturbed, such as by cutting, sanding, drilling, or building remodelling. Removal of asbestos-containing materials is not always optimal because the fibers can be spread into the air during the removal process. A management program for intact asbestos-containing materials is often recommended instead.

When asbestos-containing material is damaged or disintegrates, microscopic fibers are dispersed into the air. Inhalation of asbestos fibers over long exposure times is associated with increased incidence of lung cancer, in particular the specific form mesothelioma. The risk of lung cancer from inhaling asbestos fibers is also greater to smokers. The symptoms of the disease do not usually appear until about 20 to 30 years after the first exposure to asbestos.

Asbestos is found in older homes and buildings, but occurs most commonly in schools and industrial settings. The US Federal Government (www.osha.gov) and some states have set standards for acceptable levels of asbestos fibers in indoor air. There are particularly stringent regulations applicable to schools.

Carbon Dioxide

Carbon dioxide (CO_2) is a relatively easy to measure surrogate for indoor pollutants emitted by humans, and correlates with human metabolic activity. Carbon dioxide at levels that are unusually high indoors may cause occupants to grow drowsy, to get headaches, or to function at lower activity levels. Humans are the main indoor source of carbon dioxide in most buildings. Indoor CO_2 levels are an indicator of the adequacy of outdoor air ventilation relative to indoor occupant density and metabolic activity.

To eliminate most complaints, the total indoor CO_2 level should be reduced to a difference of less than 600 ppm above outdoor levels. The National Institute for Occupational Safety and Health (NIOSH) considers that indoor air concentrations of carbon dioxide that exceed 1,000 ppm are a marker suggesting inadequate ventilation. The UK standards for schools say that carbon dioxide in all teaching and learning spaces, when measured at seated head height and averaged over the whole day should not exceed 1,500 ppm. The whole day refers to normal school hours (i.e. 9:00am to 3:30pm) and includes unoccupied periods such as lunch breaks. In Hong Kong, the EPD established indoor air quality objectives for office buildings and public places in which a carbon dioxide level below 1,000 ppm is considered to be good. European standards limit carbon dioxide to 3,500 ppm. OSHA limits carbon dioxide concentration in the workplace to 5,000 ppm for prolonged periods, and 35,000 ppm for 15 minutes. These higher limits are concerned with avoiding loss of consciousness (fainting), and do not address impaired cognitive performance and energy, which begin to occur at lower concentrations of carbon dioxide.

Carbon dioxide concentrations increase as a result of human occupancy, but lag in time behind cumulative occupancy and intake of fresh air. The lower the air exchange rate, the slower the buildup of carbon dioxide to quasi "steady state" concentrations on which the NIOSH and UK guidance are based. Therefore, measurements of carbon dioxide for purposes of assessing the adequacy of ventilation need to be made after an extended period of steady occupancy and ventilation - in schools at least 2 hours, and in offices at least 3 hours - for concentrations to be a reasonable indicator of ventilation adequacy. Portable instruments used to measure carbon dioxide should be calibrated frequently, and outdoor measurements used for calculations should be made close in time to indoor measurements. Corrections for temperature effects on measurements made outdoors may also be necessary.

CO_2 levels in an enclosed office room can increase to over 1,000 ppm within 45 minutes.

Carbon dioxide concentrations in closed or confined rooms can increase to 1,000 ppm within 45 minutes of enclosure. For example, in a 3.5-by-4-metre (11 ft × 13 ft) sized office, atmospheric carbon dioxide increased from 500 ppm to over 1,000 ppm within 45 minutes of ventilation cessation and closure of windows and doors.

Ozone

Ozone is produced by ultraviolet light from the Sun hitting the Earth's atmosphere (especially in the ozone layer), lightning, certain high-voltage electric devices (such as air ionizers), and as a by-product of other types of pollution.

Ozone exists in greater concentrations at altitudes commonly flown by passenger jets. Reactions between ozone and onboard substances, including skin oils and cosmetics, can produce toxic chemicals as by-products. Ozone itself is also irritating to lung tissue and harmful to human health. Larger jets have ozone filters to reduce the cabin concentration to safer and more comfortable levels.

Outdoor air used for ventilation may have sufficient ozone to react with common indoor pollutants as well as skin oils and other common indoor air chemicals or surfaces. Particular concern is warranted when using "green" cleaning products based on citrus or terpene extracts, because these chemicals react very quickly with ozone to form toxic and irritating chemicals as well as fine and ultrafine particles. Ventilation with outdoor air containing elevated ozone concentrations may complicate remediation attempts.

Prompt Cognitive Deficits

In 2015, experimental studies reported the detection of significant episodic (situational) cognitive impairment from impurities in the air breathed by test subjects who were not informed about changes in the air quality. Researchers at the Harvard University and SUNY Upstate Medical University and Syracuse University measured the cognitive performance of 24 participants in three different controlled laboratory atmospheres that simulated those found in "conventional" and "green" buildings, as well as green buildings with enhanced ventilation. Performance was evaluated objectively using the widely used Strategic Management Simulation software simulation tool, which is a well-validated assessment test for executive decision-making in an unconstrained situation allowing initiative and improvisation. Significant deficits were observed in the performance scores achieved in increasing concentrations of either volatile organic compounds (VOCs) or carbon dioxide, while keeping other factors constant. The highest impurity levels reached are not uncommon in some classroom or office environments.

Effect of Indoor Plants

Spider plants (Chlorophytum comosum) absorb some airborne contaminants

Houseplants together with the medium in which they are grown can reduce components of indoor air pollution, particularly volatile organic compounds (VOC) such as benzene, toluene, and xylene. Plants remove CO_2 and release oxygen and water, although the quantitative impact for house plants is small. Most of the effect is attributed to the growing medium alone, but even this effect has finite limits associated with the type and quantity of medium and the flow of air through the medium. The effect of house plants on VOC concentrations was investigated in one study, done in a static chamber, by NASA for possible use in space colonies. The results showed that the removal of the challenge chemicals was roughly equivalent to that provided by the ventilation that occurred in a very energy efficient dwelling with a very low ventilation rate, an air exchange rate of about 1/10 per hour. Therefore, air leakage in most homes, and in non-residential buildings too, will generally remove the chemicals faster than the researchers reported for the plants tested by NASA. The most effective household plants reportedly included aloe vera, English ivy, and Boston fern for removing chemicals and biological compounds.

Plants also appear to reduce airborne microbes, molds, and increase humidity. However, the increased humidity can itself lead to increased levels of mold and even VOCs.

When CO_2 concentrations are elevated indoors relative to outdoor concentrations, it is only an indicator that ventilation is inadequate to remove metabolic products associated with human occupancy. Plants require CO_2 to grow and release oxygen when they consume CO_2. A study published in the journal Environmental Science & Technology considered uptake rates of ketones and aldehydes by the peace lily (Spathiphyllum clevelandii) and golden pothos (Epipremnum aureum.) Akira Tani and C. Nicholas Hewitt found "Longer-term fumigation results revealed that the total uptake amounts were 30–100 times as much as the amounts dissolved in the leaf, suggesting that volatile organic carbons are metabolized in the leaf and/or translocated through the petiole." It is worth noting the researchers sealed the plants in Teflon bags. "No VOC loss was detected from the bag when the plants were absent. However, when the plants were in the bag, the levels of aldehydes and ketones both decreased slowly but continuously, indicating removal by the plants". Studies done in sealed bags do not faithfully reproduce the conditions in the indoor environments of interest. Dynamic conditions with outdoor air ventilation and the processes related to the surfaces of the building itself and its contents as well as the occupants need to be studied.

While results do indicate house plants may be effective at removing some VOCs from air supplies, a review of studies between 1989 and 2006 on the performance of houseplants as air cleaners, presented at the Healthy Buildings 2009 conference in Syracuse, NY, concluded "...indoor plants have little, if any, benefit for removing indoor air of VOC in residential and commercial buildings."

Since high humidity is associated with increased mold growth, allergic responses, and respiratory responses, the presence of additional moisture from houseplants may not be desirable in all indoor settings.

HVAC Design

Environmentally sustainable design concepts also include aspects related to the commercial and residential heating, ventilation and air-conditioning (HVAC) industry. Among several considerations, one of the topics attended to is the issue of indoor air quality throughout the design and construction stages of a building's life.

One technique to reduce energy consumption while maintaining adequate air quality, is demand controlled ventilation. Instead of setting throughput at a fixed air replacement rate, carbon dioxide sensors are used to control the rate dynamically, based on the emissions of actual building occupants.

For the past several years, there have been many debates among indoor air quality specialists about the proper definition of indoor air quality and specifically what constitutes "acceptable" indoor air quality.

One way of quantitatively ensuring the health of indoor air is by the frequency of effective turnover of interior air by replacement with outside air. In the UK, for example, classrooms are required to have 2.5 outdoor air changes per hour. In halls, gym, dining, and physiotherapy spaces, the ventilation should be sufficient to limit carbon dioxide to 1,500 ppm. In the USA, and according to ASHRAE Standards, ventilation in classrooms is based on the amount of outdoor air per occupant plus the amount of outdoor air per unit of floor area, not air changes per hour. Since carbon dioxide indoors comes from occupants and outdoor air, the adequacy of ventilation per occupant is indicated by the concentration indoors minus the concentration outdoors. The value of 615 ppm above the outdoor concentration indicates approximately 15 cubic feet per minute of outdoor air per adult occupant doing sedentary office work where outdoor air contains 385 ppm, the current global average atmospheric CO_2 concentration. In classrooms, the requirements in the ASHRAE standard 62.1, Ventilation for Acceptable Indoor Air Quality, would typically result in about 3 air changes per hour, depending on the occupant density. Of course the occupants aren't the only source of pollutants, so outdoor air ventilation may need to be higher when unusual or strong sources of pollution exist indoors. When outdoor air is polluted, then bringing in more outdoor air can actually worsen the overall quality of the indoor air and exacerbate some occupant symptoms related to outdoor air pollution. Generally, outdoor country air is better than indoor city air. Exhaust gas leakages can occur from furnace metal exhaust pipes that lead to the chimney when there are leaks in the pipe and the pipe gas flow area diameter has been reduced.

The use of air filters can trap some of the air pollutants. The Department of Energy's Energy Efficiency and Renewable Energy section wrote "[Air] Filtration should have a Minimum Efficiency Reporting Value (MERV) of 13 as determined by ASHRAE 52.2-1999." Air filters are used to reduce the amount of dust that reaches the wet coils. Dust can serve as food to grow molds on the wet coils and ducts and can reduce the efficiency of the coils.

Moisture management and humidity control requires operating HVAC systems as designed. Moisture management and humidity control may conflict with efforts to try to optimize the operation to conserve energy. For example, Moisture management and humidity control requires systems to be set to supply Make Up Air at lower temperatures (design levels), instead of the higher temperatures sometimes used to conserve energy in cooling-dominated climate conditions. However, for most of the US and many parts of Europe and Japan, during the majority of hours of the year, outdoor air temperatures are cool enough that the air does not need further cooling to provide thermal comfort indoors. However, high humidity outdoors creates the need for careful attention to humidity levels indoors. High humidities give rise to mold growth and moisture indoors is associated with a higher prevalence of occupant respiratory problems.

The "dew point temperature" is an absolute measure of the moisture in air. Some facilities are being designed with the design dew points in the lower 50s °F, and some in the upper and lower 40s °F. Some facilities are being designed using desiccant wheels with gas fired heater to dry out the wheel enough to get the required dew points. On those systems, after the moisture is removed from the make up air, a cooling coil is used to lower the temperature to the desired level.

Commercial buildings, and sometimes residential, are often kept under slightly positive air pressure relative to the outdoors to reduce infiltration. Limiting infiltration helps with moisture management and humidity control.

Dilution of indoor pollutants with outdoor air is effective to the extent that outdoor air is free of harmful pollutants. Ozone in outdoor air occurs indoors at reduced concentrations because ozone is highly reactive with many chemicals found indoors. The products of the reactions between ozone and many common indoor pollutants include organic compounds that may be more odorous, irritating, or toxic than those from which they are formed. These products of ozone chemistry include formaldehyde, higher molecular weight aldehydes, acidic aerosols, and fine and ultrafine particles, among others. The higher the outdoor ventilation rate, the higher the indoor ozone concentration and the more likely the reactions will occur, but even at low levels, the reactions will take place. This suggests that ozone should be removed from ventilation air, especially in areas where outdoor ozone levels are frequently high. Recent research has shown that mortality and morbidity increase in the general population during periods of higher outdoor ozone and that the threshold for this effect is around 20 parts per billion (ppb).

Building Ecology

It is common to assume that buildings are simply inanimate physical entities, relatively stable over time. This implies that there is little interaction between the triad of the building, what is in it (occupants and contents), and what is around it (the larger environment). We commonly see the overwhelming majority of the mass of material in a building as relatively unchanged physical material over time. In fact, the true nature of buildings can be viewed as the result of a complex set of dynamic interactions among their physical, chemical, and biological dimensions. Buildings can be described and understood as complex systems. Research applying the approaches ecologists use to the understanding of ecosystems can help increase our understanding. "Building ecology " is proposed here as the application of those approaches to the built environment considering the dynamic system of buildings, their occupants, and the larger environment.

Buildings constantly evolve as a result of the changes in the environment around them as well as the occupants, materials, and activities within them. The various surfaces and the air inside a building are constantly interacting, and this interaction results in changes in each. For example, we may see a window as changing slightly over time as it becomes dirty, then is cleaned, accumulates dirt again, is cleaned again, and so on through its life. In fact, the "dirt" we see may be evolving as a result of the interactions among the moisture, chemicals, and biological materials found there.

Buildings are designed or intended to respond actively to some of these changes in and around them with heating, cooling, ventilating, air cleaning or illuminating systems. We clean, sanitize, and maintain surfaces to enhance their appearance, performance, or longevity. In other cases, such changes subtly or even dramatically alter buildings in ways that may be important to their own integrity or their impact on building occupants through the evolution of the physical, chemical, and biological processes that define them at any time. We may find it useful to combine the tools of the physical sciences with those of the biological sciences and, especially, some of the approaches used by scientists studying ecosystems, in order to gain an enhanced understanding of the environments in which we spend the majority of our time, our buildings.

Building ecology was first described by Hal Levin in an article in the April 1981 issue of Progressive Architecture magazine. A longer discussion of Building ecology can be found at and extensive resources can be found on the Building Ecology web site Building ecology.com.

Institutional Programs

The topic of IAQ has become popular due to the greater awareness of health problems caused by mold and triggers to asthma and allergies. In the US, awareness has also been increased by the involvement of the United States Environmental Protection Agency, who have developed an "IAQ Tools for Schools" program to help improve the indoor environmental conditions in educational institutions. The National Institute for Occupational Safety and Health conducts Health Hazard Evaluations (HHEs) in workplaces at the request of employees, authorised representative of employees, or employers, to determine whether any substance normally found in the place of employment has potentially toxic effects, including indoor air quality.

A variety of scientists work in the field of indoor air quality including chemists, physicists, mechanical engineers, biologists, bacteriologists and computer scientists. Some of these professionals are certified by organisations such as the American Industrial Hygiene Association, the American Indoor Air Quality Council and the Indoor Environmental Air Quality Council.

On the international level, the International Society of Indoor Air Quality and Climate (ISIAQ), formed in 1991, organises two major conferences, the Indoor Air and the Healthy Buildings series. ISIAQ's journal Indoor Air is published 6 times a year and contains peer-reviewed scientific papers with an emphasis on interdisciplinary studies including exposure measurements, modeling, and health outcomes.

Wind Rose

A wind rose is a graphic tool used by meteorologists to give a succinct view of how wind speed and direction are typically distributed at a particular location. Historically, wind roses were predecessors of the compass rose (found on maps), as there was no differentiation between a cardinal direction and the wind which blew from such a direction. Using a polar coordinate system of gridding, the frequency of winds over a time period is plotted by wind direction, with color bands showing wind speed ranges. The direction of the longest spoke shows the wind direction with the greatest frequency.

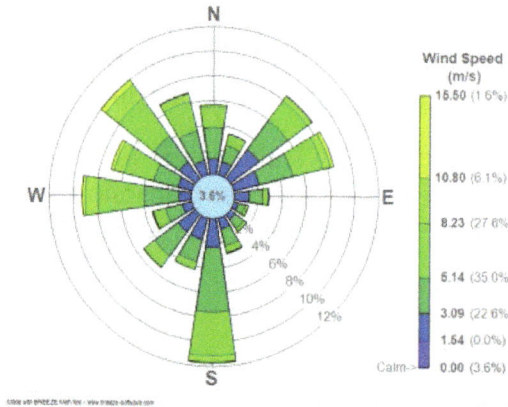

Wind rose plot for LaGuardia Airport (LGA), New York, New York. 2008

History

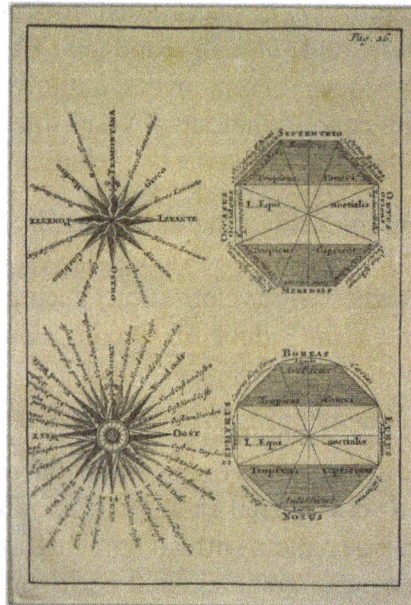

A medieval wind rose

Before the development of the compass rose, a wind rose was included on maps in order to let the reader know which directions the 8 major winds (and sometimes 8 half winds and 16 quarter winds) blew within the plan view. No differentiation was made between cardinal directions and the winds which blew from said directions. North was depicted with a fleur de lis, while east was shown as a Christian cross to indicate the direction of Jerusalem from Europe.

Use

Presented in a circular format, the modern wind rose shows the frequency of winds blowing *from* particular directions over a specified period. The length of each "spoke" around the circle is related to the frequency that the wind blows from a particular direction per unit time. Each concentric circle represents a different frequency, emanating from zero at the center to increasing frequencies at the outer circles. A wind rose plot may contain additional information, in that each spoke is broken down into color-coded bands that show wind speed ranges. Wind roses typically use 16 cardinal directions, such as north (N), NNE, NE, etc., although they may be subdivided into as many as 32 directions. In terms of angle measurement in degrees, North corresponds to 0°/360°, East to 90°, South to 180° and West to 270°.

Compiling a wind rose is one of the preliminary steps taken in constructing airport runways, as aircraft typically perform their best take-offs and landings pointing into the wind.

Continuous Emissions Monitoring System

Continuous emission monitoring systems (CEMS) were historically used as a tool to monitor flue gas for oxygen, carbon monoxide and carbon dioxide to provide information for combustion control in industrial settings. They are currently[when?] used as a means to comply with air emission standards such as the United States Environmental Protection Agency's Acid Rain Program, other federal emission programs, or state permitted emission standards. Facilities employ the use of CEMS to continuously collect, record and report the required emissions data.

The standard CEM system consists of a sample probe, filter, sample line (umbilical), gas conditioning system, calibration gas system, and a series of gas analyzers which reflect the parameters being monitored. Typical monitored emissions include: sulfur dioxide, nitrogen oxides, carbon monoxide, carbon dioxide, hydrogen chloride, airborne particulate matter, mercury, volatile organic compounds, and oxygen. CEM systems can also measure air flow, flue gas opacity and moisture.

In the U.S., the EPA requires a data acquisition and handling system to collect and report the data. SO2 emissions must be measured in pounds per hour using both an SO2 pollutant concentration monitor and a volumetric flow monitor. For NO_x, both a NO_x pollutant concentration monitor and a diluent gas monitor are required to determine the emissions rate (lbs/mmBtu). Opacity must also be monitored. NO_x measuring is not a current requirement, however if monitored, a CO_2 or oxygen monitor plus a flow monitor should be used. In monitoring these emissions, the system must be in continuous operation and must be able to sample, analyze, and record data at least every 15 minutes and then averaged hourly.

Operation

A small sample of flue gas is extracted, by means of a pump, into the CEM system via a sample probe. Facilities that combust fossil fuels often use a dilution-extractive probe to dilute the sample with clean, dry air to a ratio typically between 50:1 to 200:1, but usually 100:1. Dilution is used because pure flue gas can be hot, wet and, with some pollutants, sticky. Once diluted to the appropriate ratio, the sample is transported through a sample line (typically referred to as an umbilical) to a manifold from which individual analyzers may extract a sample. Gas analyzers employ various techniques to accurately measure concentrations. Some commonly used techniques include: infrared and ultraviolet adsorption, chemiluminescence, fluorescence and beta ray absorption. After analysis, the gas exits the analyzer to a common manifold to all analyzers where it is vented out of doors. A Data Acquisition and Handling System (DAHS) receives the signal output from each analyzer in order to collect and record emissions data.

Another sample extraction method used in industrial sources and utility sources with low emission rates, is commonly referred to as the "hot dry" extractive method or "direct" CEMS. The sample is not diluted, but is carried along a heated sample line at high temperature into a sample conditioning unit. The sample is filtered to remove particulate matter and dried, usually with a chiller, to remove moisture. Once conditioned, the sample enters a sampling manifold and is measured using the same methods above. One advantage of this method is the ability to measure % oxygen in the sample, which is often required in the regulatory calculations for emission corrections. Since dilution mixes clean dry air with the sample, dilution systems cannot measure % oxygen.

Quality Assurance

Accuracy of the system is demonstrated in several ways. An internal quality assurance check is achieved by daily introduction of a certified concentration of gas to the sample probe. The EPA also allows for the use of Continuous Emissions Monitoring Calibration Systems which dilute gases to generate calibration standards. The analyzer reading must be accurate to a certain percentage. The percent accuracy can vary, but most fall between 2.5% and 5%. In power stations affected by the Acid Rain Program, annual (or bi-annual) certification of the system must be performed by an independent firm. The firm will have an independent CEM system temporarily in place to collect emissions data in parallel with the plant CEMS. This testing is referred to as a Relative Accuracy Test Audit (RATA).

In the U.S., periodic evaluations of the equipment must be reported and recorded. This includes daily calibration error tests, daily interference tests for flow monitors, and semi-annual (or annual) RATA and bias tests. CEMS equipment is expensive and not always affordable for a facility. In such cases, a facility will install non-EPA compliant analysis equipment at the emissions point. Once yearly, for the equipment evaluation, a mobile CEMS company measures emissions with compliant equipment. The results are then compared to the non-compliant analyzer system.

Atmospheric Dispersion Modeling

Atmospheric dispersion modeling is the mathematical simulation of how air pollutants disperse in the ambient atmosphere. It is performed with computer programs that solve the mathematical equations and algorithms which simulate the pollutant dispersion. The dispersion models are used to estimate the downwind ambient concentration of air pollutants or toxins emitted from sources such as industrial plants, vehicular traffic or accidental chemical releases. They can also be used to predict future concentrations under specific scenarios (i.e. changes in emission sources). Therefore, they are the dominant type of model used in air quality policy making. They are most useful for pollutants that are dispersed over large distances and that may react in the atmosphere. For pollutants that have a very high spatio-temporal variability (i.e. have very steep distance to source decay such as black carbon) and for epidemiological studies statistical land-use regression models are also used.

Industrial air pollution source

Dispersion models are important to governmental agencies tasked with protecting and managing the ambient air quality. The models are typically employed to determine whether existing or proposed new industrial facilities are or will be in compliance with the National Ambient Air Quality Standards (NAAQS) in the United States and other nations. The models also serve to assist in the design of effective control strategies to reduce emissions of harmful air pollutants. During the late 1960s, the Air Pollution Control Office of the U.S. EPA initiated research projects that would lead to the development of models for the use by urban and transportation planners. A major and significant application of a roadway dispersion model that resulted from such research was applied to the Spadina Expressway of Canada in 1971.

Air dispersion models are also used by public safety responders and emergency management personnel for emergency planning of accidental chemical releases. Models are used to determine the consequences of accidental releases of hazardous or toxic materials, Accidental releases may result in fires, spills or explosions that involve hazardous materials, such as chemicals or radionuclides. The results of dispersion modeling, using worst case accidental release source terms and meteorological conditions, can provide an estimate of location impacted areas, ambient concentrations, and be used to determine protective actions appropriate in the event a release occurs. Appropriate protective actions may include evacuation or shelter in place for persons in the downwind direction. At industrial facilities, this type of consequence assessment or emergency planning is required under the Clean Air Act (United States) (CAA) codified in Part 68 of Title 40 of the Code of Federal Regulations.

The dispersion models vary depending on the mathematics used to develop the model, but all require the input of data that may include:

- Meteorological conditions such as wind speed and direction, the amount of atmospheric turbulence (as characterized by what is called the "stability class"), the ambient air temperature, the height to the bottom of any inversion aloft that may be present, cloud cover and solar radiation.

- Source term (the concentration or quantity of toxins in emission or accidental release source terms) and temperature of the material

- Emissions or release parameters such as source location and height, type of source (i.e., fire, pool or vent stack)and exit velocity, exit temperature and mass flow rate or release rate.

- Terrain elevations at the source location and at the receptor location(s), such as nearby homes, schools, businesses and hospitals.

- The location, height and width of any obstructions (such as buildings or other structures) in the path of the emitted gaseous plume, surface roughness or the use of a more generic parameter "rural" or "city" terrain.

Many of the modern, advanced dispersion modeling programs include a pre-processor module for the input of meteorological and other data, and many also include a post-processor module for graphing the output data and/or plotting the area impacted by the air pollutants on maps. The plots of areas impacted may also include isopleths showing areas of minimal to high concentrations that define areas of the highest health risk. The isopleths plots are useful in determining protective actions for the public and responders.

The atmospheric dispersion models are also known as atmospheric diffusion models, air dispersion models, air quality models, and air pollution dispersion models.

Atmospheric Layers

Discussion of the layers in the Earth's atmosphere is needed to understand where airborne pollutants disperse in the atmosphere. The layer closest to the Earth's surface is known as the *troposphere*. It extends from sea-level to a height of about 18 km and contains about 80 percent of the mass of the overall atmosphere. The *stratosphere* is the next layer and extends from 18 km to about 50 km. The third layer is the *mesosphere* which extends from 50 km to about 80 km. There are other layers above 80 km, but they are insignificant with respect to atmospheric dispersion modeling.

The lowest part of the troposphere is called the *atmospheric boundary layer (ABL)* or the *planetary boundary layer (PBL)* and extends from the Earth's surface to about 1.5 to 2.0 km in height. The air temperature of the atmospheric boundary layer decreases with increasing altitude until it reaches what is called the *inversion layer* (where the temperature increases with increasing altitude) that caps the atmospheric boundary layer. The upper part of the troposphere (i.e., above the inversion layer) is called the *free troposphere* and it extends up to the 18 km height of the troposphere.

The ABL is of the most important with respect to the emission, transport and dispersion of airborne pollutants. The part of the ABL between the Earth's surface and the bottom of the inversion layer is known as the mixing layer. Almost all of the airborne pollutants emitted into the ambient atmosphere are transported and dispersed within the mixing layer. Some of the emissions penetrate the inversion layer and enter the free troposphere above the ABL.

In summary, the layers of the Earth's atmosphere from the surface of the ground upwards are: the ABL made up of the mixing layer capped by the inversion layer; the free troposphere; the stratosphere; the mesosphere and others. Many atmospheric dispersion models are referred to as *boundary layer models* because they mainly model air pollutant dispersion within the ABL. To avoid confusion, models referred to as *mesoscale models* have dispersion modeling capabilities that extend horizontally up to a few hundred kilometres. It does not mean that they model dispersion in the mesosphere.

Gaussian Air Pollutant Dispersion Equation

The technical literature on air pollution dispersion is quite extensive and dates back to the 1930s and earlier. One of the early air pollutant plume dispersion equations was derived by Bosanquet and Pearson. Their equation did not assume Gaussian distribution nor did it include the effect of ground reflection of the pollutant plume.

Sir Graham Sutton derived an air pollutant plume dispersion equation in 1947 which did include the assumption of Gaussian distribution for the vertical and crosswind dispersion of the plume and also included the effect of ground reflection of the plume.

Under the stimulus provided by the advent of stringent environmental control regulations, there was an immense growth in the use of air pollutant plume dispersion calculations between the late 1960s and today. A great many computer programs for calculating the dispersion of air pollutant emissions were developed during that period of time and they were called "air dispersion models". The basis for most of those models was the Complete Equation For Gaussian Dispersion Modeling Of Continuous, Buoyant Air Pollution Plumes shown below:

$$C = \frac{Q}{u} \cdot \frac{f}{\sigma_y \sqrt{2\pi}} \cdot \frac{g_1 + g_2 + g_3}{\sigma_z \sqrt{2\pi}}$$

where:	
f	= crosswind dispersion parameter
	= $\exp[-y^2/(2\sigma_y^2)]$
g	= vertical dispersion parameter = $g_1 + g_2 + g_3$
g_1	= vertical dispersion with no reflections
	= $\exp[-(z-H)^2/(2\sigma_z^2)]$
g_2	= vertical dispersion for reflection from the ground
	= $\exp[-(z+H)^2/(2\sigma_z^2)]$
g_3	= vertical dispersion for reflection from an inversion aloft
	= $\displaystyle\sum_{m=1}^{\infty} \left\{ \exp[-(z-H-2mL)^2/(2\sigma_z^2)] \right.$
?	$+\exp[-(z-H+2mL)^2/(2\sigma_z^2)]$
?	$+\exp[-(z+H-2mL)^2/(2\sigma_z^2)]$
?	$\left. +\exp[-(z-H+2mL)^2/(2\sigma_z^2)] \right\}$

	C = concentration of emissions, in g/m³, at any receptor located:
	x meters downwind from the emission source point
	y meters crosswind from the emission plume centerline
	z meters above ground level
Q	= source pollutant emission rate, in g/s
u	= horizontal wind velocity along the plume centerline, m/s
H	= height of emission plume centerline above ground level, in m
σ_z	= vertical standard deviation of the emission distribution, in m
σ_y	= horizontal standard deviation of the emission distribution, in m
L	= height from ground level to bottom of the inversion aloft, in m
exp	= the exponential function

The above equation not only includes upward reflection from the ground, it also includes downward reflection from the bottom of any inversion lid present in the atmosphere.

The sum of the four exponential terms in g_3 converges to a final value quite rapidly. For most cases, the summation of the series with $m = 1$, $m = 2$ and $m = 3$ will provide an adequate solution.

σ_z and σ_y are functions of the atmospheric stability class (i.e., a measure of the turbulence in the ambient atmosphere) and of the downwind distance to the receptor. The two most important variables affecting the degree of pollutant emission dispersion obtained are the height of the emission source point and the degree of atmospheric turbulence. The more turbulence, the better the degree of dispersion.

The resulting calculations for air pollutant concentrations are often expressed as an air pollutant concentration contour map in order to show the spatial variation in contaminant levels over a wide area under study. In this way the contour lines can overlay sensitive receptor locations and reveal the spatial relationship of air pollutants to areas of interest.

Whereas older models rely on stability classes for the determination of σ_y and σ_z, more recent models increasingly rely on the Monin-Obukhov similarity theory to derive to derive these parameters.

Briggs Plume Rise Equations

The Gaussian air pollutant dispersion equation (discussed above) requires the input of H which is the pollutant plume's centerline height above ground level—and H is the sum of H_s (the actual physical height of the pollutant plume's emission source point) plus ΔH (the plume rise due the plume's buoyancy).

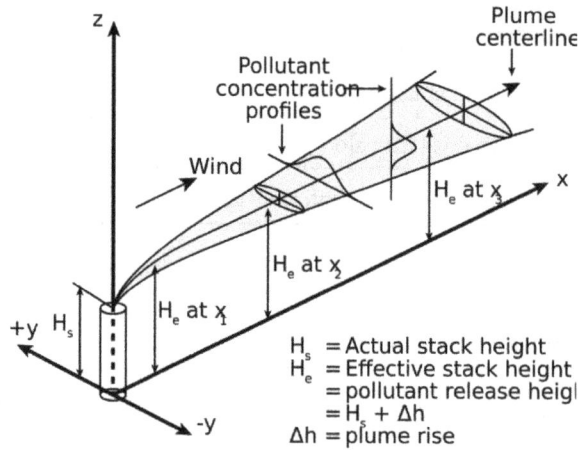

Visualization of a buoyant Gaussian air pollutant dispersion plume

To determine ΔH, many if not most of the air dispersion models developed between the late 1960s and the early 2000s used what are known as "the Briggs equations." G.A. Briggs first published his plume rise observations and comparisons in 1965. In 1968, at a symposium sponsored by CONCAWE (a Dutch organization), he compared many of the plume rise models then available in the literature. In that same year, Briggs also wrote the section of the publication edited by Slade dealing with the comparative analyses of plume rise models. That was followed in 1969 by his classical critical review of the entire plume rise literature, in which he proposed a set of plume rise equations which have become widely known as "the Briggs equations". Subsequently, Briggs modified his 1969 plume rise equations in 1971 and in 1972.

Briggs divided air pollution plumes into these four general categories:

- Cold jet plumes in calm ambient air conditions

- Cold jet plumes in windy ambient air conditions

- Hot, buoyant plumes in calm ambient air conditions

- Hot, buoyant plumes in windy ambient air conditions

Briggs considered the trajectory of cold jet plumes to be dominated by their initial velocity momentum, and the trajectory of hot, buoyant plumes to be dominated by their buoyant momentum to the extent that their initial velocity momentum was relatively unimportant. Although Briggs proposed plume rise equations for each of the above plume categories, *it is important to emphasize that "the Briggs equations" which become widely used are those that he proposed for bent-over, hot buoyant plumes.*

In general, Briggs's equations for bent-over, hot buoyant plumes are based on observations and data involving plumes from typical combustion sources such as the flue gas stacks from steam-generating boilers burning fossil fuels in large power plants. Therefore, the stack exit velocities were probably in the range of 20 to 100 ft/s (6 to 30 m/s) with exit temperatures ranging from 250 to 500 °F (120 to 260 °C).

A logic diagram for using the Briggs equations to obtain the plume rise trajectory of bent-over buoyant plumes is presented below:

LOGIC DIAGRAM FOR BRIGGS' EQUATIONS TO CALCULATE THE RISE OF A BUOYANT PLUME

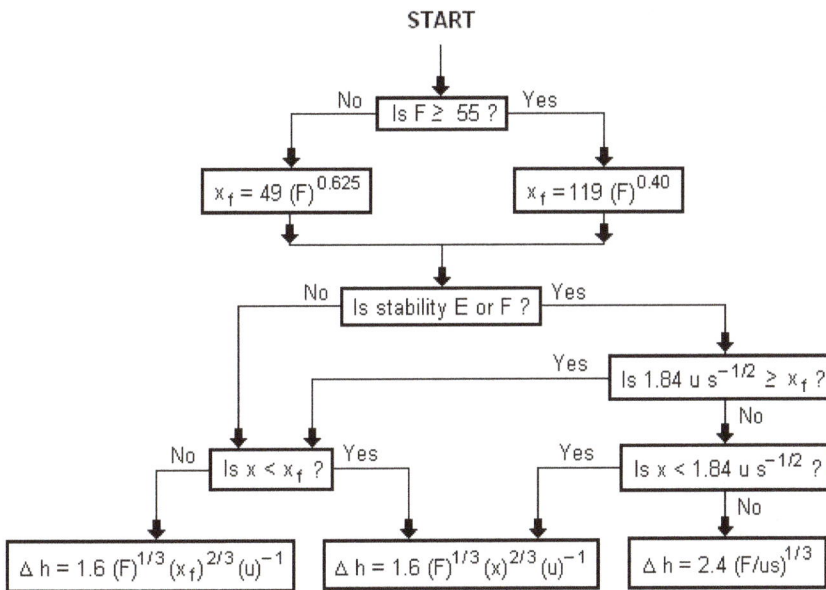

START

Is F ≥ 55 ?   No / Yes

$x_f = 49\,(F)^{0.625}$

$x_f = 119\,(F)^{0.40}$

Is stability E or F ?   No / Yes

Is $1.84\,u\,s^{-1/2} \geq x_f$?   Yes / No

Is $x < x_f$?   No / Yes

Is $x < 1.84\,u\,s^{-1/2}$?   Yes / No

$\Delta h = 1.6\,(F)^{1/3}\,(x_f)^{2/3}\,(u)^{-1}$

$\Delta h = 1.6\,(F)^{1/3}\,(x)^{2/3}\,(u)^{-1}$

$\Delta h = 2.4\,(F/us)^{1/3}$

where:	
Δh	= plume rise, in m
F	= buoyancy factor, in $m^4 s^{-3}$
x	= downwind distance from plume source, in m
x_f	= downwind distance from plume source to point of maximum plume rise, in m
u	= windspeed at actual stack height, in m/s
s	= stability parameter, in s^{-2}

The above parameters used in the Briggs' equations are discussed in Beychok's book.

References

- Garcia, Javier; Colosio, Joëlle (2002). Air-quality indices : elaboration, uses and international comparisons. Presses des MINES. ISBN 2-911762-36-3.

- The Babcock & Wilcox Company. Steam: its generation and use. The Babcock & Wilcox Company. pp. 36–5. ISBN 0-9634570-1-2.

- Turner, D.B. (1994). Workbook of atmospheric dispersion estimates: an introduction to dispersion modeling (2nd ed.). CRC Press. ISBN 1-56670-023-X.

- "Specifications and Test Procedures for Total Hydrocarbon Continuous Monitoring Systems in Stationary Sources" (PDF). www3.epa.gov. Retrieved 23 February 2016.

- National Weather Service Corporate Image Web Team. "NOAA's National Weather Service/Environmental Protection Agency - United States Air Quality Forecast Guidance". Retrieved 20 August 2015.

- Rama Lakshmi (17 October 2014). "India launches its own Air Quality Index. Can its numbers be trusted?". Washington Post. Retrieved 20 August 2015.

- "National Air Quality Index (AQI) launched by the Environment Minister AQI is a huge initiative under 'Swachh Bharat'". Retrieved 20 August 2015.

- "Air Quality Index (AQI) - A Guide to Air Quality and Your Health". US EPA. 9 December 2011. Retrieved 8 August 2012.

4

Monitoring of Water Pollution

Water pollution has become one of the major problems faced by our world in today's time. Considering how vital water is to life, it deserves our utmost attention. There are a number of ways to monitor water pollution; some of these are water quality modeling, bacteriological water analysis, wastewater quality indicators and water-sensitive urban design. Water pollution can best be understood in confluence with the major topics listed in the following section.

Water Quality

Water quality refers to the chemical, physical, biological, and radiological characteristics of water. It is a measure of the condition of water relative to the requirements of one or more biotic species and or to any human need or purpose. It is most frequently used by reference to a set of standards against which compliance can be assessed. The most common standards used to assess water quality relate to health of ecosystems, safety of human contact, and drinking water.

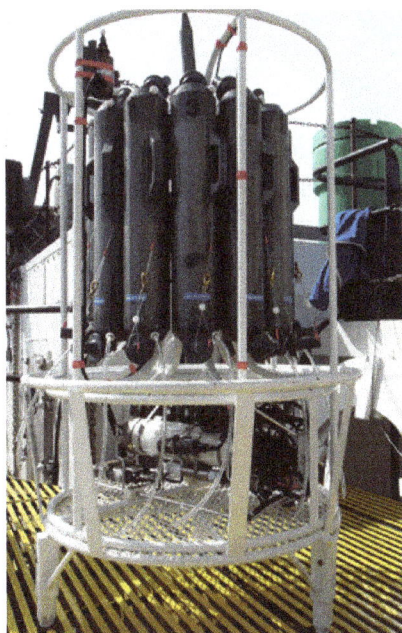

A rosette sampler is used to collect water samples in deep water, such as the Great Lakes or oceans, for water quality testing.

Standards

In the setting of standards, agencies make political and technical/scientific decisions about how the water will be used. In the case of natural water bodies, they also make some reasonable estimate of pristine conditions. Different uses raise different concerns and therefore different standards are considered. Natural water bodies will vary in response to environmental conditions. Environmental scientists work to understand how these systems function, which in turn helps to identify the sources and fates of contaminants. Environmental lawyers and policymakers work to define legislation with the intention that water is maintained at an appropriate quality for its identified use.

The vast majority of surface water on the planet is neither potable nor toxic. This remains true when seawater in the oceans (which is too salty to drink) is not counted. Another general perception of *water quality* is that of a simple property that tells whether water is polluted or not. In fact, water quality is a complex subject, in part because water is a complex medium intrinsically tied to the ecology of the Earth. Industrial and commercial activities (e.g. manufacturing, mining, construction, transport) are a major cause of water pollution as are runoff from agricultural areas, urban runoff and discharge of treated and untreated sewage.

Categories

The parameters for water quality are determined by the intended use. Work in the area of water quality tends to be focused on water that is treated for human consumption, industrial use, or in the environment.

Human Consumption

Contaminants that may be in untreated water include microorganisms such as viruses, protozoa and bacteria; inorganic contaminants such as salts and metals; organic chemical contaminants from industrial processes and petroleum use; pesticides and herbicides; and radioactive contaminants. Water quality depends on the local geology and ecosystem, as well as human uses such as sewage dispersion, industrial pollution, use of water bodies as a heat sink, and overuse (which may lower the level of the water).

The United States Environmental Protection Agency (EPA) limits the amounts of certain contaminants in tap water provided by US public water systems. The Safe Drinking Water Act authorizes EPA to issue two types of standards: *primary standards* regulate substances that potentially affect human health, and *secondary standards* prescribe aesthetic qualities, those that affect taste, odor, or appearance. The U.S. Food and Drug Administration (FDA) regulations establish limits for contaminants in bottled water that must provide the same protection for public health. Drinking water, including bottled water, may reasonably be expected to contain at least small amounts of some contaminants. The presence of these contaminants does not necessarily indicate that the water poses a health risk.

In urbanized areas around the world, water purification technology is used in municipal water systems to remove contaminants from the source water (surface water or groundwater) before it is distributed to homes, businesses, schools and other recipients. Water drawn directly from a stream, lake, or aquifer and that has no treatment will be of uncertain quality.

Industrial and Domestic Use

Dissolved minerals may affect suitability of water for a range of industrial and domestic purposes. The most familiar of these is probably the presence of ions of calcium and magnesium which interfere with the cleaning action of soap, and can form hard sulfate and soft carbonate deposits in water heaters or boilers. Hard water may be softened to remove these ions. The softening process often substitutes sodium cations. Hard water may be preferable to soft water for human consumption, since health problems have been associated with excess sodium and with calcium and magnesium deficiencies. Softening decreases nutrition and may increase cleaning effectiveness.

Environmental Water Quality

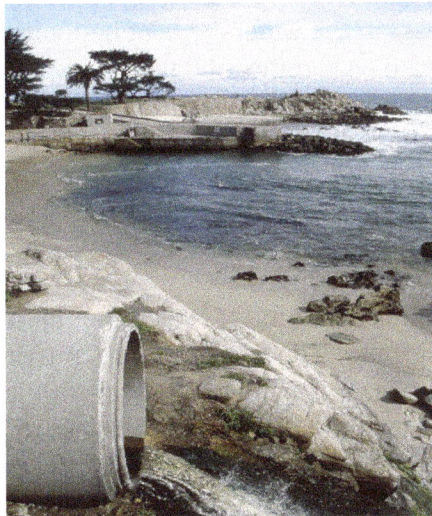

Urban runoff discharging to coastal waters

Environmental water quality, also called ambient water quality, relates to water bodies such as lakes, rivers, and oceans. Water quality standards for surface waters vary significantly due to different environmental conditions, ecosystems, and intended human uses. Toxic substances and high populations of certain microorganisms can present a health hazard for non-drinking purposes such as irrigation, swimming, fishing, rafting, boating, and industrial uses. These conditions may also affect wildlife, which use the water for drinking or as a habitat. Modern water quality laws generally specify protection of fisheries and recreational use and require, as a minimum, retention of current quality standards.

Satirical cartoon by William Heath, showing a woman observing monsters in a drop of London water (at the time of the *Commission on the London Water Supply* report, 1828)

There is some desire among the public to return water bodies to pristine, or pre-industrial conditions. Most current environmental laws focus on the designation of particular uses of a water body. In some countries these designations allow for some water contamination as long as the particular type of contamination is not harmful to the designated uses. Given the landscape changes (e.g., land development, urbanization, clearcutting in forested areas) in the watersheds of many freshwater bodies, returning to pristine conditions would be a significant challenge. In these cases, environmental scientists focus on achieving goals for maintaining healthy ecosystems and may concentrate on the protection of populations of endangered species and protecting human health.

Sampling and Measurement

The complexity of water quality as a subject is reflected in the many types of measurements of water quality indicators. The most accurate measurements of water quality are made on-site, because water exists in equilibrium with its surroundings. Measurements commonly made on-site and in direct contact with the water source in question include temperature, pH, dissolved oxygen, conductivity, oxygen reduction potential (ORP), turbidity, and Secchi disk depth.

Sample Collection

The second problem occurs as the sample is removed from the water source and begins to establish chemical equilibrium with its new surroundings - the sample container. Sample containers must be made of materials with minimal reactivity with substances to be measured; and pre-cleaning of sample containers is important. The water sample may dissolve part of the sample container and any residue on that container, or chemicals dissolved in the water sample may sorb onto the sample container and re-main there when the water is poured out for analysis. Similar physical and chemical interactions may take place with any pumps, piping, or intermediate devices used to transfer the water sample into the sample container. Water collected from depths below the surface will normally be held at the reduced pressure of the atmosphere; so gas dissolved in the water may escape into unfilled space at the top of the container. Atmospheric gas present in that air space may also dissolve into the water sample. Other chemical reaction equilibria may change if the water sample changes temperature. Finely divided solid particles formerly suspended by water turbulence may settle to the bottom of the sample container, or a solid phase may form from biological growth or chemical precipitation. Microorganisms within the water sample may biochemically alter concentrations of oxygen, carbon dioxide, and organic compounds. Changing carbon dioxide concentrations may alter pH and change solubility of chemicals of interest. These problems are of special concern during measurement of chemicals assumed to be significant at very low concentrations.

An automated sampling station installed along the East Branch Milwaukee River, New Fane, Wisconsin. The cover of the 24-bottle autosampler (center) is partially raised, showing the sample bottles inside. The autosampler was programmed to collect samples at time intervals, or proportionate to flow over a specified period. The data logger (white cabinet) recorded temperature, specific conductance, and dissolved oxygen levels.

Sample preservation may partially resolve the second problem. A common procedure is keeping samples cold to slow the rate of chemical reactions and phase change, and analyzing the sample as soon as possible; but this merely minimizes the changes rather than preventing them. A useful procedure for determining influence of sample containers during delay between sample collection and analysis involves preparation for two artificial samples in advance of the sampling event. One sample container is filled with water known from previous analysis to contain no detectable amount of the chemical of interest. This sample, called a "blank," is opened for exposure to the atmosphere when the sample of interest is collected, then resealed and transported to the laboratory with the sample for analysis to determine if sample holding procedures introduced any measurable amount of the chemical of interest. The second artificial sample is collected with the sample of interest, but then "spiked" with a measured additional amount of the chemical of interest at the time of collection. The blank and spiked samples are carried with the sample of interest and analyzed by the same methods at the same times to determine any changes indicating gains or losses during the elapsed time between collection and analysis.

Filtering a manually collected water sample (grab sample) for analysis

Testing in Response to Natural Disasters and Other Emergencies

Inevitably after events such as earthquakes and tsunamis, there is an immediate response by the aid agencies as relief operations get underway to try and restore basic infrastructure and provide the basic fundamental items that are necessary for survival and subsequent recovery. Access to clean drinking water and adequate sanitation is a priority at times like this. The threat of disease increases hugely due to the large numbers of people living close together, often in squalid conditions, and without proper sanitation.

After a natural disaster, as far as water quality testing is concerned there are widespread views on the best course of action to take and a variety of methods can be employed. The key basic water quality parameters that need to be addressed in an emergency are bacteriological indicators of fecal contamination, free chlorine residual, pH, turbidity and possibly conductivity/total dissolved solids. There are a number of portable water test kits on the market widely used by aid and relief agencies for carrying out such testing.

After major natural disasters, a considerable length of time might pass before water quality returns to pre-disaster levels. For example, following the 2004 Indian Ocean Tsunami the Colombo-based International Water Management Institute (IWMI) monitored the effects of saltwater and concluded that the wells recovered to pre-tsunami drinking water quality one and a half years after the event. IWMI developed protocols for cleaning wells contaminated by saltwater; these were subsequently officially endorsed by the World Health Organization as part of its series of Emergency Guidelines.

Chemical Analysis

The simplest methods of chemical analysis are those measuring chemical elements without respect to their form. Elemental analysis for oxygen, as an example, would indicate a concentration of 890,000 milligrams per litre (mg/L) of water sample because water is made of oxygen. The method selected to measure dissolved oxygen should differentiate between diatomic oxygen and oxygen combined with other elements. The comparative simplicity of elemental analysis has produced a large amount of sample data and water quality criteria for elements sometimes identified as heavy metals. Water analysis for heavy metals must consider soil particles suspended in the water sample. These suspended soil particles may contain measurable amounts of metal. Although the particles are not dissolved in the water, they may be consumed by people drinking the water. Adding acid to a water sample to prevent loss of dissolved metals onto the sample container may dissolve more metals from suspended soil particles. Filtration of soil particles from the water sample before acid addition, however, may cause loss of dissolved metals onto the filter. The complexities of differentiating similar organic molecules are even more challenging.

A gas chromatograph-mass spectrometer measures
pesticides and other organic pollutants

Making these complex measurements can be expensive. Because direct measurements of water quality can be expensive, ongoing monitoring programs are typically conducted by government agencies. However, there are local volunteer programs and resources available for some general assessment. Tools available to the general public include on-site test kits, commonly used for home fish tanks, and biological assessment procedures.

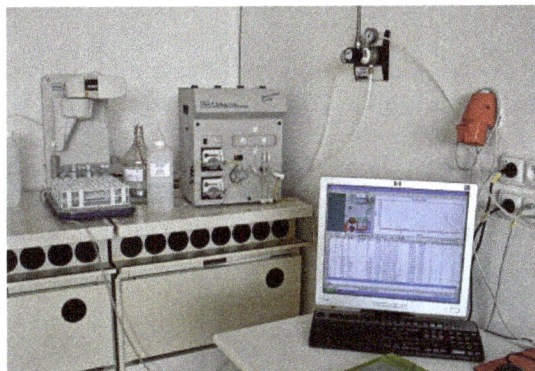

Atomic fluorescence spectroscopy is used to measure mercury and other heavy metals

Real-time Monitoring

Although water quality is usually sampled and analyzed at laboratories, nowadays, citizens demand real-time information about the water they are drinking. During the last years, several companies are deploying worldwide real-time remote monitoring systems for measuring water pH, turbidity or dissolved oxygen levels.

Drinking Water Indicators

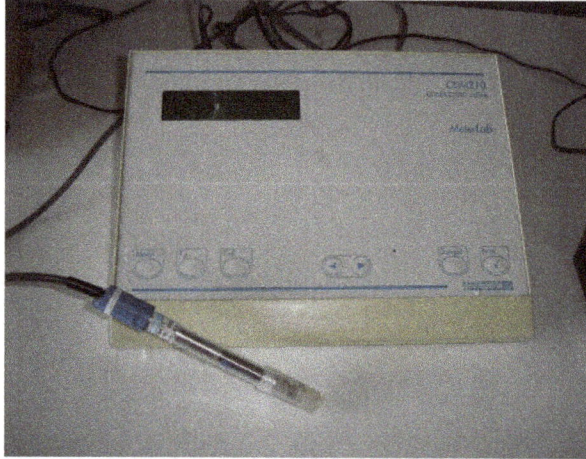

An electrical conductivity meter is used to measure total dissolved solids

The following is a list of indicators often measured by situational category:

- Alkalinity

- Color of water

- pH

- Taste and odor (geosmin, 2-Methylisoborneol (MIB), etc.)

- Dissolved metals and salts (sodium, chloride, potassium, calcium, manganese, magnesium)

- Microorganisms such as fecal coliform bacteria (*Escherichia coli*), Cryptosporidium, and Giardia lamblia

- Dissolved metals and metalloids (lead, mercury, arsenic, etc.)

- Dissolved organics: colored dissolved organic matter (CDOM), dissolved organic carbon (DOC)

- Radon

- Heavy metals

- Pharmaceuticals

- Hormone analogs

Environmental Indicators

Physical Indicators

• Water Temperature	• Total dissolved solids (TDS)
• Specifics Conductance or EC, Electrical Conductance, Conductivity	• Odour of water
	• Color of water
• Total suspended solids (TSS)	• Taste of water
• Transparency or Turbidity	

Chemical Indicators

• pH	• Heavy metals
• Biochemical oxygen demand (BOD)	• Nitrate
	• Orthophosphates
• Chemical oxygen demand (COD)	• Pesticides
• Dissolved oxygen (DO)	• Surfactants
• Total hardness (TH)	

Biological Indicators

• Ephemeroptera	• *Escherichia coli* (E. coli)
• Plecoptera	
• Mollusca	• Coliform bacteria
• Trichoptera	

Biological monitoring metrics have been developed in many places, and one widely used measure is the presence and abundance of members of the insect orders Ephemeroptera, Plecoptera and Trichoptera. (Common names are, respectively, Mayfly, Stonefly and Caddisfly.) EPT indexes will naturally vary from region to region, but generally, within a region, the greater the number of taxa from these orders, the better the water quality. Organisations in the United States, such as EPA offer guidance on developing a monitoring program and identifying members of these and other aquatic insect orders.

Individuals interested in monitoring water quality who cannot afford or manage lab scale analysis can also use biological indicators to get a general reading of water quality. One example is the IOWATER volunteer water monitoring program, which includes a benthic macroinvertebrate indicator key.

Bivalve molluscs are largely used as bioindicators to monitor the health of aquatic environments in both fresh water and the marine environments. Their population status or structure, physiology, behaviour or the level of contamination with elements or compounds can indicate the state of contamination status of the ecosystem. They are particularly useful since they are sessile so that they are representative of the environment where they are sampled or placed. A typical project is the Mussel Watch Programme, but today they are used worldwide.

The Southern African Scoring System (SASS) method is a biological water quality monitoring system based on the presence of benthic macroinvertebrates. The SASS aquatic biomonitoring tool has been refined over the past 30 years and is now on the fifth version (SASS5) which has been specifically modified in accordance with international standards, namely the ISO/IEC 17025 protocol. The SASS5 method is used by the South African Department of Water Affairs as a standard method for River Health Assessment, which feeds the national River Health Programme and the national Rivers Database.

Standards and Reports

International

- The World Health Organisation (WHO) has published guidelines for drinking-water quality (GDWQ) in 2011.

- The International Organization for Standardization (ISO) published[when?] regulation of water quality in the section of ICS 13.060, ranging from water sampling, drinking water, industrial class water, sewage, and examination of water for chemical, physical or biological properties. ICS 91.140.60 covers the standards of water supply systems.

National Specifications for Ambient Water and Drinking Water

European Union

The water policy of the European Union is primarily codified in three directives:

- Directive on Urban Waste Water Treatment (91/271/EEC) of 21 May 1991 concerning discharges of municipal and some industrial wastewaters;

- The Drinking Water Directive (98/83/EC) of 3 November 1998 concerning potable water quality;

- Water Framework Directive (2000/60/EC) of 23 October 2000 concerning water resources management.

India

- Indian Council of Medical Research (ICMR) Standards for Drinking Water.

South Africa

Water quality guidelines for South Africa are grouped according to potential user types (e.g. domestic, industrial) in the 1996 Water Quality Guidelines. Drinking water quality is subject to the South African National Standard (SANS) 241 Drinking Water Specification.

United Kingdom

In England and Wales acceptable levels for drinking water supply are listed in the "Water Supply (Water Quality) Regulations 2000."

United States

In the United States, Water Quality Standards are defined by state agencies for various water bodies, guided by the desired uses for the water body (e.g., fish habitat, drinking water supply, recreational use). The Clean Water Act (CWA) requires each governing jurisdiction (states, territories, and covered tribal entities) to submit a set of biennial reports on the quality of water in their area. These reports are known as the 303(d) and 305(b) reports, named for their respective CWA provisions, and are submitted to, and approved by, EPA. These reports are completed by the governing jurisdiction, typically a state environmental agency. EPA recommends that each state submit a single "Integrated Report" comprising its list of impaired waters and the status of all water bodies in the state. The *National Water Quality Inventory Report to Congress* is a general report on water quality, providing overall information about the number of miles of streams and rivers and their aggregate condition. The CWA requires states to adopt standards for each of the possible designated uses that they assign to their waters. Should evidence suggest or document that a stream, river or lake has failed to meet the water quality criteria for one or more of its designated uses, it is placed on a list of impaired waters. Once a state has placed a water body on this list, it must develop a management plan establishing Total Maximum Daily Loads (TMDLs) for the pollutant(s) impairing the use of the water. These TMDLs establish the reductions needed to fully support the designated uses.

Drinking water standards, which are applicable to public water systems, are issued by EPA under the Safe Drinking Water Act.

Water Quality Modelling

Water quality modeling involves the prediction of water pollution using mathematical simulation techniques. A typical water quality model consists of a collection of formulations representing physical mechanisms that determine position and momentum of pollutants in a water body. Models are available for individual components of the hydrological system such as surface runoff; there also exist basinwide models addressing hydrologic transport and for ocean and estuarine applications. Often finite difference methods are used to analyse these phenomena, and, almost always, large complex computer models are required.

Formulations and Associated Constants

Water quality is modeled by one or more of the following formulations:

- Advective Transport formulation

- Dispersive Transport formulation

- Surface Heat Budget formulation

- Dissolved Oxygen Saturation formulation

- Reaeration formulation

- Carbonaceous Deoxygenation formulation

- Nitrogenous Biochemical Oxygen Demand formulation

- Sediment oxygen demand formulation (SOD)

- Photosynthesis and Respiration formulation

- pH and Alkalinity formulation

- Nutrients formulation (fertilizers)

- Algae formulation

- Zooplankton formulation

- Coliform bacteria formulation (e.g. *Escherichia coli*)

Bacteriological Water Analysis

Bacteriological water analysis is a method of analysing water to estimate the numbers of bacteria present and, if needed, to find out what sort of bacteria they are. It represents one aspect of water quality. It is a microbiological analytical procedure which uses samples of water and from these samples determines the concentration of bacteria. It is then possible to draw inferences about the suitability of the water for use from these concentrations. This process is used, for example, to routinely confirm that water is safe for human consumption or that bathing and recreational waters are safe to use.

The interpretation and the action trigger levels for different waters vary depending on the use made of the water. Very stringent levels applying to drinking water whilst more relaxed levels apply to marine bathing waters, where much lower volumes of water are expected to be ingested by users.

Approach

The common feature of all these routine screening procedures is that the primary analysis is for indicator organisms rather than the pathogens that might cause concern. Indicator organisms are bacteria such as non-specific coliforms, *Escherichia coli* and *Pseudomonas aeruginosa* that are very commonly found in the human or animal gut and which, if detected, may suggest the presence of sewage. Indicator organisms are used because even when a person is infected with a more pathogenic bacteria, they will still be excreting many millions times more indicator organisms than pathogens. It is therefore reasonable to surmise that if indicator organism levels are low, then pathogen levels will be very much lower or absent. Judgements as to suitability of water for use are based on very extensive precedents and relate to the probability of any sample population of bacteria being able to be infective at a reasonable statistical level of confidence.

Analysis is usually performed using culture, biochemical and sometimes optical methods. When indicator organisms levels exceed pre-set triggers, specific analysis for pathogens may then be undertaken and these can be quickly detected (where suspected) using specific culture methods or molecular biology.

Methodologies

Because the analysis is always based on a very small sample taken from a very large volume of water, all methods rely on statistical principles.

Multiple Tube Method

One of the oldest methods is called the multiple tube method. In this method a measured sub-sample (perhaps 10 ml) is diluted with 100 ml of sterile growth medium and an aliquot of 10 ml is then decanted into each of ten tubes. The remaining 10 ml is then diluted again and the process repeated. At the end of 5 dilutions this produces 50 tubes covering the dilution range of 1:10 through to 1:10000.

The tubes are then incubated at a pre-set temperature for a specified time and at the end of the process the number of tubes with growth in is counted for each dilution.

Statistical tables are then used to derive the concentration of organisms in the original sample. This method can be enhanced by using indicator medium which changes colour when acid forming species are present and by including a tiny inverted tube called a Durham tube in each sample tube. The Durham inverted tube catches any gas produced. The production of gas at 37 degrees Celsius is a strong indication of the presence of *Escherichia coli.*

ATP Testing

An ATP test is the process of rapidly measuring active microorganisms in water through detection adenosine triphosphate (ATP). ATP is a molecule found only in and around living cells, and as such it gives a direct measure of biological concentration and health. ATP is quantified by measuring the light produced through its reaction with the naturally-occurring enzyme firefly luciferase using a luminometer. The amount of light produced is directly proportional to the amount of biological energy present in the sample.

Second generation ATP tests are specifically designed for water, wastewater and industrial applications where, for the most part, samples contain a variety of components that can interfere with the ATP assay.

Plate Count

The plate count method relies on bacteria growing a colony on a nutrient medium so that the colony becomes visible to the naked eye and the number of colonies on a plate can be counted. To be effective, the dilution of the original sample must be arranged so that on average between 30 and 300 colonies of the target bacterium are grown. Fewer than 30 colonies makes the interpretation statistically unsound whilst greater than 300 colonies often results in overlapping colonies and imprecision in the count. To ensure that an appropriate number of colonies will be generated several dilutions are normally cultured. This approach is widely utilised for the evaluation of the effectiveness of water treatment by the inactivation of representative microbial contaminants such as "E. coli" following ASTM D5465.

The laboratory procedure involves making serial dilutions of the sample (1:10, 1:100, 1:1000, etc.) in sterile water and cultivating these on nutrient agar in a dish that is sealed and incubated. Typical media include plate count agar for a general count or MacConkey agar to count Gram-negative bacteria such as *E. coli.* Typically one set of plates is incubated at 22°C and for 24 hours and a second set at 37°C for 24 hours. The composition of the nutrient usually includes reagents that resist the growth of non-target organisms and make the target organism easily identified, often by a colour change in the medium. Some recent methods include a fluorescent agent so that counting of the colonies can be automated. At the end of the incubation period the colonies are counted by eye, a procedure that takes a few moments and does not require a microscope as the colonies are typically a few millimetres across.

Membrane Filtration

Most modern laboratories use a refinement of total plate count in which serial dilutions of the sample are vacuum filtered through purpose made membrane filters and these filters are themselves laid on nutrient medium within sealed plates. The methodology is otherwise similar to conventional total plate counts. Membranes have a printed millimetre grid printed on and can be reliably used to count the number of colonies under a binocular microscope.

Pour Plate Method

When the analysis is looking for bacterial species that grow poorly in air, the initial analysis is done by mixing serial dilutions of the sample in liquid nutrient agar which is then poured into bottles which are then sealed and laid on their sides to produce a sloping agar surface. Colonies that develop in the body of the medium can be counted by eye after incubation.

The total number of colonies is referred to as the Total Viable Count (TVC). The unit of measurement is cfu/ml (or colony forming units per millilitre) and relates to the original sample. Calculation of this is a multiple of the counted number of colonies multiplied by the dilution used.

Pathogen Analysis

When samples show elevated levels of indicator bacteria, further analysis is often undertaken to look for specific pathogenic bacteria. Species commonly investigated in the temperate zone include *Salmonella typhi* and *Salmonella Typhimurium*. Depending on the likely source of contamination investigation may also extend to organisms such as *Cryptosporidium spp*. In tropical areas analysis of *Vibrio cholerae* is also routinely undertaken.

Types of Nutrient Media Used in Analysis

MacConkey agar is culture medium designed to grow Gram-negative bacteria and stain them for lactose fermentation. It contains bile salts (to inhibit most Gram-positive bacteria), crystal violet dye (which also inhibits certain Gram-positive bacteria), neutral red dye (which stains microbes fermenting lactose), lactose and peptone. Alfred Theodore MacConkey developed it while working as a bacteriologist for the Royal Commission on Sewage Disposal in the United Kingdom.

Endo agar contains peptone, lactose, dipotassium phosphate, agar, sodium sulfite, basic fuchsin and was originally developed for the isolation of *Salmonella typhi*, but is now commonly used in water analysis. As in MacConkey agar, coliform organisms ferment the lactose, and the colonies become red. Non-lactose-fermenting organisms produce clear, colourless colonies against the faint pink background of the medium.

mFC medium is used in membrane filtration and contains selective and differential agents. These include Rosolic acid to inhibit bacterial growth in general, except for faecal coliforms, Bile salts inhibit non-enteric bacteria and Aniline blue indicates the ability of faecal coliforms to ferment lactose to acid that causes a pH change in the medium.

TYEA medium contains tryptone, yeast extract, common salt and L-arabinose per liter of glass distilled water and is a non selective medium usually cultivated at two temperatures (22 and 36°C) to determine a general level of contamination (a.k.a. colony count).

Wastewater Quality Indicators

Wastewater quality indicators are laboratory test methodologies to assess suitability of wastewater for disposal or re-use. Tests selected and desired test results vary with the intended use or discharge location. Tests measure physical, chemical, and biological characteristics of the waste water.

Physical Characteristics

Temperature

Aquatic organisms cannot survive outside of specific temperature ranges. Irrigation runoff and water cooling of power stations may elevate temperatures above the acceptable range for some species. Temperature may be measured with a calibrated thermometer.

Solids

Solid material in wastewater may be dissolved, suspended, or settled. Total dissolved solids or TDS (sometimes called filterable residue) is measured as the mass of residue remaining when a measured volume of filtered water is evaporated. The mass of dried solids remaining on the filter is called total suspended solids (TSS) or nonfiltrable residue. Settleable solids are measured as the visible volume accumulated at the bottom of an Imhoff cone after water has settled for one hour. Turbidity is a measure of the light scattering ability of suspended matter in the water. Salinity measures water density or conductivity changes caused by dissolved materials.

Chemical Characteristics

Virtually any chemical may be found in water, but routine testing is commonly limited to a few chemical elements of unique significance.

Hydrogen

Water ionizes into hydronium (H_3O) cations and hydroxyl (OH) anions. The concentration of ionized hydrogen (as protonated water) is expressed as pH.

Oxygen

Most aquatic habitats are occupied by fish or other animals requiring certain minimum dissolved oxygen concentrations to survive. Dissolved oxygen concentrations may be measured directly in wastewater, but the amount of oxygen potentially required by other chemicals in the wastewater is termed an oxygen demand. Dissolved or suspended oxidizable organic material in wastewater will be used as a food source. Finely divided material is readily available to microorganisms whose populations will increase to digest the amount of food available. Digestion of this food requires oxygen, so the oxygen content of the water will ultimately be decreased by the amount required to digest the dissolved or suspended food. Oxygen concentrations may fall below the minimum required by aquatic animals if the rate of oxygen utilization exceeds replacement by atmospheric oxygen.

The reaction for biochemical oxidation may be written as:

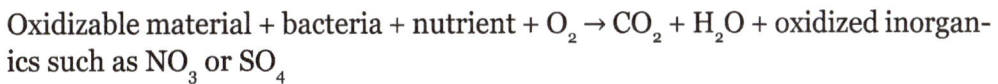

Oxidizable material + bacteria + nutrient + O_2 → CO_2 + H_2O + oxidized inorganics such as NO_3 or SO_4

Oxygen consumption by reducing chemicals such as sulfides and nitrites is typified as follows:

$$S^{2-} + 2\,O_2 \rightarrow SO_4^{2-}$$

$$NO_2^- + \tfrac{1}{2}\,O_2 \rightarrow NO_3^-$$

Since all natural waterways contain bacteria and nutrient, almost any waste compounds introduced into such waterways will initiate biochemical reactions (such as shown above). Those biochemical reactions create what is measured in the laboratory as the biochemical oxygen demand (BOD).

Oxidizable chemicals (such as reducing chemicals) introduced into a natural water will similarly initiate chemical reactions (such as shown above). Those chemical reactions create what is measured in the laboratory as the chemical oxygen demand (COD).

Both the BOD and COD tests are a measure of the relative oxygen-depletion effect of a waste contaminant. Both have been widely adopted as a measure of pollution effect. The BOD test measures the oxygen demand of biodegradable pollutants whereas the COD test measures the oxygen demand of biodegradable pollutants plus the oxygen demand of non-biodegradable oxidizable pollutants.

The so-called 5-day BOD measures the amount of oxygen consumed by biochemical oxidation of waste contaminants in a 5-day period. The total amount of oxygen consumed when the biochemical reaction is allowed to proceed to completion is called the Ultimate BOD. The Ultimate BOD is too time consuming, so the 5-day BOD has almost universally been adopted as a measure of relative pollution effect.

There are also many different COD tests. Perhaps, the most common is the 4-hour COD.

There is no generalized correlation between the 5-day BOD and the Ultimate BOD. Likewise, there is no generalized correlation between BOD and COD. It is possible to develop such correlations for a specific waste contaminant in a specific wastewater stream, but such correlations cannot be generalized for use with any other waste contaminants or wastewater streams.

The laboratory test procedures for determining the above oxygen demands are detailed in the following sections of the "Standard Methods For the Examination Of Water and Wastewater" available at www.standardmethods.org:

- 5-day BOD and Ultimate BOD: Sections 5210B and 5210C

- COD: Section 5220

Nitrogen

Nitrogen is an important nutrient for plant and animal growth. Atmospheric nitrogen is less biologically available than dissolved nitrogen in the form of ammonia and nitrates. Availability of dissolved nitrogen may contribute to algal blooms. Ammonia and organic forms of nitrogen are often measured as Total Kjeldahl Nitrogen, and analysis for inorganic forms of nitrogen may be performed for more accurate estimates of total nitrogen content.

Phosphates

Total Phosphorus and Phosphate, $PO-34$

Phosphates enter the water ways through both non-point sources and point sources. Non-point source (NPS) pollution refers to water pollution from diffuse sources. Non-point source pollution can be contrasted with point source pollution, where discharges occur to a body of water at a single location. The non-point sources of phosphates include: natural decomposition of rocks and minerals, storm water runoff, agricultural runoff, erosion and sedimentation, atmospheric deposition, and direct input by animals/wildlife; whereas: point sources may include: waste water treatment plants and permitted industrial discharges. In general, the non-point source pollution typically is significantly higher than the point sources of pollution. Therefore, the key to sound management is to limit the input from both point and non-point sources of phosphate. High concentration of phosphate in water bodies is an indication of pollution and largely responsible for eutrophication.

Phosphates are not toxic to people or animals unless they are present in very high levels. Digestive problems could occur from extremely high levels of phosphate.

The following criteria for total phosphorus were recommended by the U.S. Environmental Protection Agency.

1. No more than 0.1 mg/L for streams which do not empty into reservoirs,

2. No more than 0.05 mg/L for streams discharging into reservoirs, and

3. No more than 0.025 mg/L for reservoirs.

Phosphorus is normally low (< 1 mg/l) in clean potable water sources and usually not regulated;

Chlorine

Chlorine has been widely used for bleaching, as a disinfectant, and for biofouling prevention in water cooling systems. Remaining concentrations of oxidizing hypochlorous acid and hypochlorite ions may be measured as chlorine residual to estimate effectiveness of disinfection or to demonstrate safety for discharge to aquatic ecosystems.

Biological Characteristics

Water may be tested by a bioassay comparing survival of an aquatic test species in the wastewater in comparison to water from some other source. Water may also be evaluated to determine the approximate biological population of the wastewater. Pathogenic micro-organisms using water as a means of moving from one host to another may be present in sewage. Coliform index measures the population of an organism commonly found in the intestines of warm-blooded animals as an indicator of the possible presence of other intestinal pathogens.

Freshwater Environmental Quality Parameters

Freshwater environmental quality parameters are the natural and man-made chemical, biological and microbiological characteristics of rivers, lakes and ground-waters, the ways they are measured and the ways that they change. The values or concentrations attributed to such parameters can be used to describe the pollution status of an environment, its biotic status or to predict the likelihood or otherwise of a particular organisms being present. Monitoring of environmental quality parameters is a key activity in managing the environment, restoring polluted environments and anticipating the effects of man-made changes on the environment.

Freshwater environmental quality parameters are those chemical, physical or biological parameters that can be used to characterise a freshwater body. Because almost all water bodies are dynamic in their composition, the relevant quality parameters are typically expressed as a range of expected concentrations.

Characterisation

The first step in understanding the chemistry of freshwater is to establish the relevant concentrations of the parameters of interest. Conventionally this is done by taking representative samples of the water for subsequent analysis in a laboratory . However, in-situ monitoring using hand-held analytical equipment or using bank-side monitoring stations are also used.

Sampling

Freshwaters are surprisingly difficult to sample because they are rarely homogeneous and their quality varies during the day and during the year. In addition the most representative sampling locations are often at a distance from the shore or bank increasing the logistic complexity.

Rivers

Filling a clean bottle with river water is a very simple task, but a single sample is only representative of that point along the river the sample was taken from and at that point in time. Understanding the chemistry of a whole river, or even a significant tributary, requires prior investigative work to understand how homogeneous or mixed the flow is and to determine if the quality changes during the course of a day and during the course of a year. Almost all natural rivers will have very significant patterns of change through the day and through the seasons. Many rivers also have a very large flow that is unseen. This flows through underlying gravel and sand layers and is called the hyporheic zone How much mixing there is between the hyporheic zone and the water in the open channel will depend on a variety of factors, some of which relate to flows leaving aquifers which may have been storing water for many years.

Ground-waters

Ground waters by their very nature are often very difficult to access to take a sample. As a consequence the majority of ground-water data comes from samples taken from springs, wells, water supply bore-holes and in natural caves. In recent decades as the need to understand ground water dynamics has increased, an increasing number or monitoring bore-holes have been drilled into aquifers

Lakes

Lakes and ponds can be very large and support a complex eco-system in which environmental parameters vary widely in all three physical dimensions and with time. Large lakes in the temperate zone often stratify in the warmer months into a warmer upper layers rich in oxygen and a colder lower layer with low oxygen levels. In the autumn, falling temperatures and occasional high winds result in the mixing of the two layers into a more homogeneous whole. When stratification occurs it not only affects oxygen levels but also many related parameters such as iron, phosphate and manganese which are all changed in their chemical form by change in the redox potential of the environment.

Lakes also receive waters, often from many different sources with varying qualities. Solids from stream inputs will typically settle near the mouth of the stream and depending on a variety of factors the incoming water may float over the surface of the lake, sink beneath the surface or rapidly mix with the lake water. All of these phenomena can skew the results of any environmental monitoring unless the process are well understood.

Mixing Zones

Where two rivers meet at a confluence there exists a mixing zone. A mixing zone may be very large and extend for many miles as in the case of the Mississippi and Missouri rivers in the United States and the River Clwyd and River Elwy in North Wales. In a mixing zone water chemistry may be very variable and can be difficult to predict. The chemical interactions are not just simple mixing but may be complicated by biological processes from submerged macrophytes and by water joining the channel from the hyporheic zone or from springs draining an aquifer.

Geological Inputs

The geology that underlies a river or lake has a major impact on its chemistry. A river flowing across very ancient precambrian schists is likely to have dissolved very little from the rocks and maybe similar to de-ionised water at least in the headwaters. Conversely a river flowing through chalk hills, and especially if its source is in the chalk, will have a high concentration of carbonates and bicarbonates of Calcium and possibly Magnesium.

As a river progresses along its course it may pass through a variety of geological types and it may have inputs from aquifers that do not appear on the surface anywhere in the locality.

Atmospheric Inputs

Oxygen is probably the most important chemical constituent of surface water chemistry, as all aerobic organisms require it for survival. It enters the water mostly via diffusion at the water-air interface. Oxygen's solubility in water decreases as water temperature increases. Fast, turbulent streams expose more of the water's surface area to the air and tend to have low temperatures and thus more oxygen than slow, backwaters. Oxygen is a by-product of photosynthesis, so systems with a high abundance of aquatic algae and plants may also have high concentrations of oxygen during the day. These levels can decrease significantly during the night when primary producers switch to respiration. Oxygen can be limiting if circulation between the surface and deeper layers is poor, if the activity of animals is very high, or if there is a large amount of organic decay occurring such as following Autumn leaf-fall.

Most other atmospheric inputs come from man-made or anthropogenic sources the most significant of which are the oxides of sulphur produced by burning sulphur rich fuels such as coal and oil which give rise to acid rain. The chemistry of sulphur oxides is complex both in the atmosphere and in river systems. However the effect on the overall chemistry is simple in that it reduces the pH of the water making it more acidic. The pH change is most marked in rivers with very low concentrations of dissolved salts as these cannot buffer the effects of the acid input. Rivers downstream of major industrial conurbations are also at greatest risk. In parts of Scandinavia and West Wales and Scotland many rivers became so acidic from oxides of sulphur that most fish life was destroyed and pHs as low as pH4 were recorded during critical weather conditions.

Anthropogenic Inputs

The majority of rivers on the planet and many lakes have received or are receiving inputs from human-kind's activities. In the industrialised world, many rivers have been very seriously polluted, at least during the 19th and the first half of the 20th centuries. Although in general there has been much improvement in the developed world, there is still a great deal of river pollution apparent on the planet.

Toxicity

In most environmental situations the presence or absence of an organism is determined by a complex web of interactions only some of which will be related to measurable chemical or biological parameters. Flow rate, turbulence, inter and intra specific competition, feeding behaviour, disease, parasatism, commensalism and symbiosis are just a few of the pressures and opportunities facing any organism or population. Most chemical constituents favour some organisms and are less favourable to others. However, there are some cases where a chemical constituent exerts a toxic effect. i.e. where the concentration can kill or severely inhibit the normal functioning of the organism. Where a toxic effect has been demonstrated this may be noted in the sections below dealing with the individual parameters.

Chemical Constituents

Colour and Turbidity

Often it is the colour of freshwater or how clear or hazy the water is that is the most obvious visual characteristic. Unfortunately neither colour nor turbidity are strong indicators of the overall chemical composition of water. However both colour and turbidity reduce the amount of light penetrating the water and can have significant impact on algae and macrophytes. Some algae in particular are highly dependent on water with low colour and turbidity

Many rivers draining high moor-lands overlain by peat have a very deep yellow brown colour caused by dissolved humic acids.

Organic Constituents

One of the principal sources of elevated concentrations of organic chemical constituents is from treated sewage.

Dissolved organic material is most commonly measured using either the Biochemical oxygen demand (BOD) test or the Chemical oxygen demand (COD) test. Organic constituents are significant in river chemistry for the effect that they have on dissolved oxygen concentration and for the impact that individual organic species may have directly on aquatic biota.

Any organic and degradable material consumes oxygen as it decomposes. Where organic concentrations are significantly elevated the effects on oxygen concentrations can be significant and as conditions get extreme the river bed may become anoxic.

Some organic constituents such as synthetic hormones, pesticides, phthalates have direct metabolic effects on aquatic biota and even on humans drinking water taken from the river. Understanding such constituents and how they can be identified and quantified is becoming of increasing importance in the understanding of freshwater chemistry.

Metals

A wide range of metals may be found in rivers from natural sources where metal ores are present in the rocks over which the river flows or in the aquifers feeding water into the river. However many rivers have an increased load of metals because of industrial activities which include mining and quarrying and the processing and use of metals.

Iron

Iron, usually as Fe^{+++} is a common constituent of river waters at very low levels. Higher iron concentrations in acidic springs or an anoxic hyporheic zone may cause visible orange/brown staining or semi-gelatinous precipitates of dense orange iron bacterial floc carpeting the river bed. Such conditions are very deleterious to most organisms and can cause serious damage in a river system.

Coal mining is also a very significant source of Iron both in mine-waters and from stocking yards of coal and from coal processing. Long abandoned mines can be a highly intractable source of high concentrations of Iron. Low levels of iron are common in spring waters emanating from deep-seated aquifers and maybe regarding as health giving springs. Such springs are commonly called Chalybeate springs and have given rise to a number of Spa towns in Europe and the United States.

Zinc

Zinc is normally associated with metal mining, especially Lead and Silver mining but is also a component pollutant associated with a variety of other metal mining activities and with Coal mining. Zinc is toxic at relatively low concentrations to many aquatic organisms. *Microregma* starts to show a toxic reaction at concentrations as low as 0.33 mg/l

Heavy Metals

Lead and silver in river waters are commonly found together and associated with lead mining. Impacts from very old mines can be very long-lived. In the River Ystwyth in Wales for example, the effects of silver and lead mining in the 17th and 18th centuries in the headwaters still causes unacceptably high levels of Zinc and Lead in the river water right down to its confluence with the sea. Silver is very toxic even at very low concentrations but leaves no visible evidence of its contamination.

Lead is also highly toxic to freshwater organisms and to humans if the water is used as drinking water. As with Silver, Lead pollution is not visible to the naked eye. The River Rheidol in west Wales had a major series of lead mines in its headwaters until the end of the 19th century and its mine discharges and waste tips remain to this day. In 1919 - 1921 only 14 species of invertebrates were found in the lower Rheidol when Lead concentrations were between 0.2ppm and 0.5ppm. By 1932 the lead concentration had reduced to 0.02ppm to 0.1ppm because of the abandonment of mining and, at those concentrations, the bottom fauna had stabilized to 103 species including three leeches.

Coal mining is also a very significant source of metals, especially Iron, Zinc and Nickel particularly where the coal is rich if pyrites which oxidises on contact with the air producing a very acidic leachate which is able to dissolve metals from the coal.

Significant levels of copper are unusual in rivers and where it does it occur the source is most likely to be mining activities, coal stocking, or pig farming. Rarely elevated levels may be of geological origin. Copper is acutely toxic to many freshwater organisms, especially algae, at very low concentrations and significant concentration in river water may have serious adverse effects on the local ecology.

Nitrogen

Nitrogenous compounds have a variety of sources including washout of oxides of nitrogen from the atmosphere, some geological inputs and some from macrophyte and algal nitrogen fixation. However, for many rivers in the proximity of humans, the largest input is from sewage whether treated or untreated. The nitrogen derives from breakdown products of proteins found in urine and faeces. These products, being very soluble, often pass through sewage treatment process and are discharged into rivers as a component of sewage treatment effluent. Nitrogen may be in the form of nitrate, nitrite, ammonia or ammonium salts or what is termed albuminoid nitrogen or nitrogen still within an organic proteinoid molecule.

The differing forms of nitrogen are relatively stable in most river systems with nitrite slowly transforming into nitrate in well oxygenated rivers and ammonia transforming into nitrite/ nitrate. However, the process are slow in cool rivers and reduction in concentration may more often be attributed to simple dilution. All forms of nitrogen are taken up by macrophytes and algae and elevated levels of nitrogen are often associated with overgrowths of plants or eutrophication. These can have the effect of blocking channels and inhibiting navigation. However, ecologically, the more significant effect is on dissolved oxygen concentrations which may become super-saturated during daylight due to plant photosynthesis but then drop to very low levels during darkness as plant respiration uses up the dissolved oxygen. Coupled with the release of oxygen in photosynthesis is the creation of bi-carbonate ions which cause a steep rise in pH and this is matched in darkness as carbon dioxide is released through respiration which substantially lowers the pH. Thus high levels of nitrogenous compounds tends to lead to eutrophication with extreme variations in parameters which in turn can substantially degrade the ecological worth of the watercourse.

Ammonium ions also have a toxic effect, especially on fish. The toxicity of ammonia is dependent on both pH and temperature and an added complexity is the buffering effect of the blood/water interface across the gill membrane which masks any additional toxicity over about pH 8.0. The management of river chemistry to avoid ecological damage is particularly difficult in the case of ammonia as a wide range of potential scenarios of concentration, pH and temperature have to be considered and the diurnal pH fluctuation caused by photosynthesis considered. On warm summer days with high-bi-carbonate concentrations unexpectedly toxic conditions can be created.

Phosphorus

Phosphorus compounds are usually found as relatively insoluble phosphates in river water and, except in some exceptional circumstances, their origin is agriculture or human sewage. Phosphorus can encourage excessive growths of plants and algae and contribute to eutrophication. If a river discharges into a lake or reservoir phosphate can be mobilised year after year by natural processes. In the summer time, lakes stratify so that warm oxygen rich water floats on top of cold oxygen poor water. In the warm upper layers - the epilimnion- plants consume the available phosphate. As the plants die in the late summer they fall into the cool water layers underneath - the hypolimnion - and decompose. During winter turn-over, when a lake becomes fully mixed through the action of winds on a cooling body of water - the phosphates are spread throughout the lake again to feed a new generation of plants. This process is one of the principal causes of persistent algal blooms at some lakes.

Arsenic

Geological deposits of arsenic may be released into rivers where deep ground-waters are exploited as in parts of Pakistan. Many metalloid ores such as lead, gold and copper contain traces of arsenic and poorly stored tailings may result in arsenic entering the hydrological cycle.

Solids

Inert solids are produced in all montane rivers as the energy of the water helps grind away rocks into gravel, sand and finer material. Much of this settles very quickly and provides an important substrate for many aquatic organisms. Many salmonid fish require beds of gravel and sand in which to lay their eggs. Many other types of solids from agriculture, mining, quarrying, urban run-off and sewage may block-out sunlight from the river and may block interstices in gravel beds making them useless for spawning and supporting insect life.

Bacterial, Viral and Parasite Inputs

Both agriculture and sewage treatment produce inputs into rivers with very high concentrations of bateria and viruses including a wide range of pathogenic organisms. Even in areas with little human activity significant levels of bacteria and viruses can be detected originating from fish and aquatic mammals and from animals grazing near rivers such as deer. Upland waters draining areas frequented by sheep, goats or deer may also harbour a variety of opportunistic human parasites such as liver fluke. Consequently, there are very few rivers from which the water is safe to drink without some form of sterilisation or disinfection. In rivers used for contact recreation such as swimming, safe levels of bacteria and viruses can be established based on risk assessment.

Under certain conditions bacteria can colonise freshwaters occasionally making large rafts of filamentous mats known as *sewage fungus* – usually *Sphaerotilus natans*. The presence of such organisms is almost always an indicator of extreme organic pollution and would be expected to be matched with low dissolved oxygen concentrations and high BOD vales.

E. coli bacteria have been commonly found in recreational waters and their presence is used to indicate the presence of recent fecal contamination, but E. coli presence may not be indicative of human waste. E. coli are harbored in all warm-blooded animals: birds and mammals alike. E. coli bacteria have also been found in fish and turtles. Sand also harbors E. coli bacteria and some strains of E. coli have become naturalized. Some geographic areas may support unique populations of E. coli and conversely, some E. coli strains are cosmopolitan .

pH

pH in rivers is affected by the geology of the water source, atmospheric inputs and a range of other chemical contaminants. pH is only likely to become an issue on very poorly buffered upland rivers where atmospheric sulphur and nitrogen oxides may very significantly depress the pH as low as pH4 or in eutrophic alkaline rivers where photosynthetic bi-carbonate ion production in photosynthesis may drive the pH up above pH10

Water-sensitive Urban Design

Water-sensitive urban design (WSUD) is a land planning and engineering design approach which integrates the urban water cycle, including stormwater, groundwater and wastewater management and water supply, into urban design to minimise environmental degradation and improve aesthetic and recreational appeal. WSUD is a term used in the Middle East and Australia and is similar to low-impact development (LID), a term used in the United States; and sustainable urban drainage systems (SUDS), a term used in the United Kingdom.

Background

Traditional urban and industrial development alters landscapes from permeable vegetated surfaces to a series of impervious interconnected surfaces resulting in large quantities of stormwater runoff, requiring management. Historically Australia, like other industrialised countries including the United States and United Kingdom, has treated stormwater runoff as a liability and nuisance endangering human health and property. This resulted in a strong focus on the design of stormwater management systems that rapidly convey stormwater runoff directly to streams with little or no focus on ecosystem preservation. This management approach results in what is referred to as urban stream syndrome. Heavy rainfall flows rapidly into streams carrying pollutants and sediments washed off from impervious surfaces, resulting in streams carrying elevated concentrations of pollutants, nutrients and suspended solids. Increased peak flow also alters channel morphology and stability, further proliferating sedimentation and drastically reducing biotic richness.

Increased recognition of urban stream syndrome in the 1960s resulted in some movement towards holistic stormwater management in Australia. Awareness increased greatly during the 1990s with the Federal government and scientists cooperating through the Cooperative Research Centre program. Increasingly city planners have recognised the need for an integrated management approach to potable, waste and stormwater management, to enable cities to adapt and become resilient to the pressure which population growth, urban densification and climate change places on ageing and increasingly expensive water infrastructure. Additionally, Australia's arid conditions means it is particularly vulnerable to climate change, which together with its reliance on surface water sources, combined with one of the most severe droughts (from 2000–2010) since European settlement, highlight the fact that major urban centres face increasing water shortages. This has begun shifting the perception of stormwater runoff from strictly a liability and nuisance to that of having value as a water resource resulting in changing stormwater management practices.

Australian states, building on the federal government's foundational research in the 1990s, began releasing WSUD guidelines with Western Australia first releasing guidelines in 1994. Victoria released guidelines on the best practice environmental management of urban stormwater in 1999 (developed in consultation with New South Wales) and similar documents were released by Queensland through Brisbane City Council in 1999. Cooperation between federal, state and territory governments to increase the efficiency of Australia's water use resulted in the National Water Initiative (NWI) signed in June 2004. The NWI is a comprehensive national strategy to improve water management across the country, encompasses a wide range of water management issues and encourages the adoption of best practice approaches to the management of water in Australia which include WSUD.

Differences to Conventional Urban Stormwater Management

WSUD regards urban stormwater runoff as a resource rather than a nuisance or liability. This represents a paradigm shift in the way environmental resources and water infrastructure are dealt with in the planning and design of towns and cities. WSUD principles regard all streams of water as a resource with diverse impacts on biodiversity, water and land and the community's recreational and aesthetic enjoyment of waterways.

Principles

- Protecting and enhancing creeks, rivers and wetlands within urban environments;

- Protecting and improving the water quality of water draining from urban environments into creeks, rivers and wetlands;

- Restoring the urban water balance by maximising the reuse of stormwater, recycled water and grey water;

- Conserving water resources through reuse and system efficiency;

- Integrating stormwater treatment into the landscape so that it offers multiple beneficial uses such as water quality treatment, wildlife habitat, recreation and open public space;

- Reducing peak flows and runoff from the urban environment simultaneously providing for infiltration and groundwater recharge;

- Integrating water into the landscape to enhance urban design as well as social, visual, cultural and ecological values; and

- Easy and cost effective implementation of WSUD allowing for widespread application.

Objectives

- Reducing potable water demand through demand and supply side water management;

- Incorporating the use of water efficient appliances and fittings;

- Adopting a fit-for-purpose approach to the use of potential alternative sources of water such as rainwater;

- Minimising wastewater generation and treatment of wastewater to a standard suitable for effluent reuse and/or release to receiving waters;

- Treating stormwater to meet water quality objectives for reuse and/or discharge by capturing sediments, pollution and nutrients through the retention and slow release of stormwater;

- Improving waterway health through restoring or preserving the natural hydrological regime of catchments through treatment and reuse technologies;

- Improving aesthetics and the connection with water for the urban dwellers;

- Promoting a significant degree of water-related self-sufficiency within urban settings by optimizing the use of water sources to minimise potable storm and waste water inflows and outflows through the incorporation into urban design of localised water storage;

- Counteracting the 'urban heat island effect' through the use of water and vegetation assisting in replenishing groundwater.

Techniques

- The use of water-efficient appliances to reduce potable water use;

- Greywater reuse as an alternate source of water to conserve potable supplies;

- Detention, rather than rapid conveyance, of stormwater;

- Reuse, storage and infiltration of stormwater, instead of drainage system augmentation;

- Use of vegetation for stormwater filtering purposes;

- Water efficient landscaping to reduce potable water consumption;

- Protection of water-related environmental, recreational and cultural values by minimising the ecological footprint of a project associated with providing supply, wastewater and stormwater services;

- Localised wastewater treatment and reuse systems to reduce potable water consumption and minimise environmentally harmful wastewater discharges;

- Provision of stormwater or other recycled urban waters (in all cases subject to appropriate controls) to provide environmental water requirements for modified watercourses;

- Flexible institutional arrangements to cope with increased uncertainty and variability in climate;

- A focus on longer term planning; and

- A diverse portfolio of water sources, supported by both centralised and decentralised water infrastructure.

Common WSUD Practices

Common WSUD practices used in Australia are discussed below. Usually, a combination of these elements are used to meet urban water cycle management objectives.

Road Layout and Streetscape

Bioretention Systems

Bioretention systems involve treatment by vegetation prior to filtration of sediment and other solids through prescribed media. Vegetation provides biological uptake of nitrogen, phosphorus and other soluble or fine particulate contaminants. Bioretention systems offer a smaller footprint than other similar measures (e.g. constructed wetlands) and are commonly used to filter and treat runoff prior to it reaching street drains. Use on larger scales can be complicated and hence other devices may be more appropriate. Biorentention systems comprise bioretention swales (also referred to as grassed swales and drainage channels) and bioretention basins.

Bioretention Swales

Bioretention swales, similar to buffer strips and swales, are placed within the base of a swale that is generally located in the median strip of divided roads. They provide both stormwater treatment and conveyance functions. A bioretention system can be installed in part of a swale, or along the full length of a swale, depending on treatment requirements. The runoff water usually goes through a fine media filter and proceeds downwards where it is collected via a perforated pipe leading to downstream waterways or storages. Vegetation growing in the filter media can prevent erosion and, unlike infiltration systems, bioretention swales are suited for a wide range of soil conditions.

Bioretention Basins

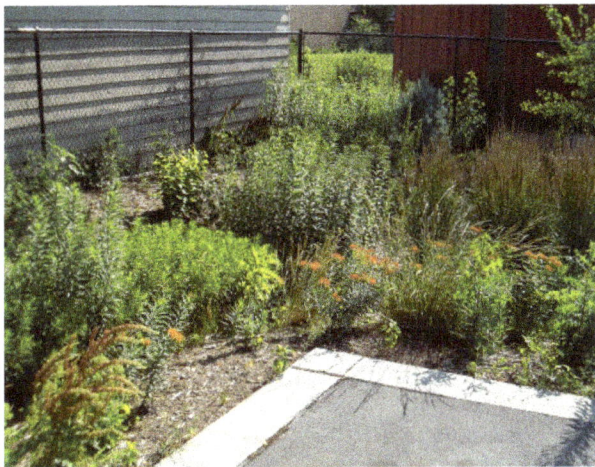

Parking lot that drains to a small bioretention basin.

Bioretention basins provide similar flow control and water quality treatment functions to bioretention swales but do not have a conveyance function. In addition to the filtration and biological uptake functions of bioretention systems, basins also provide extended detention of stormwater to maximise runoff treatment during small to medium flow events. The term raingarden is also used to describe such systems but usually refers to smaller, individual lot-scale bioretention basins. Bioretention basins have the advantage of being applicable at a range of scales and shapes and therefore have flexibility in their location within developments. Like other bioretention systems, they are often located along streets at regular intervals to treat runoff prior to entry into the drainage system. Alternatively, larger basins can provide treatment for larger areas, such as at the outfalls of a drainage system. A wide range of vegetation can be used within a bioretention basin, allowing them to be well integrated into the surrounding landscape design. Vegetation species that tolerate periodic inundation should be selected. Bioretention basins are however, sensitive to any materials that may clog the filter media. Basins are often used in conjunction with gross pollutant traps (GPTs or litter traps, include widely used trash racks), and coarser sediment basins, which capture litter and other gross solids to reduce the potential for damage to the vegetation or filter media surface.

Infiltration Trenches and Systems

Infiltration trenches are shallow excavated structures filled with permeable materials such as gravel or rock to create an underground reservoir. They are designed to hold stormwater runoff within a subsurface trench and gradually release it into the surrounding soil and groundwater systems. Although they are generally not designed as a treatment measure but can provide some level of treatment by retaining pollutants and sediments. Runoff volumes and peak discharges from impervious areas are reduced by capturing and infiltrating flows.

Due to their primary function of being the discharge of treated stormwater, infiltration systems are generally positioned as the final element in a WSUD system. Infiltration trenches should not be located on steep slopes or unstable areas. A layer of geotextile fabric is often used to line the trench in order to prevent the soil from migrating into the rock or gravel fill. Infiltration systems are dependent on the local soil characteristics and are generally best suited to soils with good infiltrative capacity, such as sandy-loam soils, with deep groundwater. In areas of low permeability soils, such as clay, a perforated pipe may be placed within the gravel.

Regular maintenance is crucial to ensure that the system does not clog with sediments and that the desired infiltration rate is maintained. This includes checking and maintaining the pre-treatment by periodic inspections and cleaning of clogged material.

Sand Filters

Sand filters are a variation of the infiltration trench principle and operate in a way similar to bioretention systems. Stormwater is passed through them for treatment prior to discharge to the downstream stormwater system. Sand filters are very useful in treating runoff from confined hard surfaces such as car parks and from heavily urbanised and built-up areas. They usually do not support vegetation owing to the filtration media (sand) not retaining sufficient moisture and because they are usually installed underground. The filter usually consists of a sedimentation chamber as pre-treatment device to remove litter, debris, gross pollutants and medium-sized sediments; a weir; followed by a sand layer that filters sediments, finer particulates and dissolved pollutants. The filtered water is collected by perforated underdrain pipes in a similar manner as in bioretention systems. Systems may also have an overflow chamber. The sedimentation chamber can have permanent water or can be designed to be drained with weep holes between storm events. Permanent water storage however, can risk anaerobic conditions that can lead to the release of pollutants (e.g. phosphorus). The design process should consider the provision of detention storage to yield a high hydrologic effectiveness, and discharge control by proper sizing of the perforated underdrain and overflow path. Regular maintenance is required to prevent crust forming.

Porous Paving

Porous paving (or pervious paving) is an alternative to conventional impermeable pavement and allows infiltration of runoff water to the soil or to a dedicated water storage reservoir below it In reasonably flat areas such as car parks, driveways and lightly used roads, it decreases the volume and velocity of stormwater runoff and can improve water quality by removing contaminants through filtering, interception and biological treatment. Porous pavements can have several forms and are either monolithic or modular. Monolithic structures consist of a single continuous porous medium such as porous concrete or porous pavement (asphalt) while modular structures include porous pavers individual paving blocks that are constructed so that there is a gap in between each paver. Commercial products that are available are for example, pavements made from special asphalt or concrete containing minimal materials, concrete grid pavements, and concrete ceramic or plastic modular pavements. Porous pavements are usually laid on a very porous material (sand or gravel), underlain by a layer of geotextile material. Maintenance activities vary depending on the type of porous pavement. Generally, inspections and removal of sediment and debris should be undertaken. Modulate pavers can also be lifted, backwashed and replaced when blockages occurs. Generally porous pavement is not suited for areas with heavy traffic loads. Particulates in stormwater can clog pores in the material.

Public Open Space

Sedimentation Basins

Sediment basin installed on a construction site.

Sedimentation basins (otherwise known as sediment basins) are used to remove (by settling) coarse to medium-sized sediments and to regulate water flows and are often the first element in a WSUD treatment system. They operate through temporary stormwater retention and reduction of flow velocities to promote settling of sediments out of the water column. They are important as a pretreatment to ensure downstream elements are not overloaded or smothered with coarse sediments. Sedimentation basins can take various forms and can be used as permanent systems integrated into an urban design or temporary measures to control sediment discharge during construction activities. They are often designed as an inlet pond to a bioretention basin or constructed wetland. Sedimentation basins are generally most effective at removing coarser sediments (125 μm and larger) and are typically designed to remove 70 to 90% of such sediments. They can be designed to drain during periods without rainfall and then fill during runoff events or to have a permanent pool. In flow events greater than their designed discharge, a secondary spillway directs water to a bypass channel or conveyance system, preventing the resuspension of sediments previously trapped in the basin.

Constructed Wetlands

Constructed wetlands are designed to remove stormwater pollutants associated with fine to colloidal particles and dissolved contaminants. These shallow, extensively vegetated water bodies use enhanced sedimentation, fine filtration and biological uptake to remove these pollutants. They usually comprise three zones: an inlet zone (sedimentation basin) to remove coarse sediments; a macrophyte zone, a heavily vegetated area to remove fine particulates and uptake of soluble pollutants; and a high flow bypass channel to protect the macrophyte zone. The macrophyte zone generally includes a marsh zone as well as an open water zone and has an extended depth of 0.25 to 0.5m with specialist plant species and a retention time of 48 to 72 hours. Constructed Wetlands can also provide a flow control function by rising during rainfall and then slowly releasing the stored flows. Constructed wetlands will improve the runoff water quality depending on the wetland processes. The key treatment mechanism of wetlands are physical (trapping suspended solids and adsorbed pollutants), biological and chemical uptake (trapping dissolved pollutants, chemical adsorption of pollutants), and pollutant transformation (more stable sediment fixation, microbial processes, UV disinfection).

The design of constructed wetlands requires careful consideration to avoid common problems such as accumulation of litter, oil and scum in sections of the wetland, infestation of weeds, mosquito problems or algal blooms. Constructed wetlands can require a large amount of land area and are unsuitable for steep terrain. High costs of the area and of vegetation establishment can be deterrents to the use of constructed wetlands as a WSUD measure. Guidelines for developers (such as the Urban Stormwater: Best Practice Environmental Management Guidelines in Victoria) require the design to retain particles of 125μm and smaller with very high efficiency and to reduce typical pollutants (such as phosphorus and nitrogen) by at least 45%. In addition to stormwater treatment, the design criteria for constructed wetlands also include enhanced aesthetic and recreational values, and habitat provision. The maintenance of constructed wetlands usually includes the removal of sediments and litter from the inlet zone, as well as weed control and occasional macrophyte harvesting to maintain a vigorous vegetation cover.

Swales And Buffer Strips

Swales and buffer strips are used to convey stormwater in lieu of pipes and provide a buffer strip between receiving waters (e.g. creek or wetland) and impervious areas of a catchment. Overland flows and mild slopes slowly convey water downstream and promote an even distribution of flow. Buffer areas provide treatment through sedimentation and interaction with vegetation.

Two swales for a housing development. The foreground one is under construction while the background one is established.

Swales can be incorporated in urban designs along streets or parklands and add to the aesthetic character of an area. Typical swales are created with longitudinal slopes between 1% and 4% in order to maintain flow capacity without creating high velocities, potential erosion of the bioretention or swale surface and safety hazard. In steeper areas check banks along swales or dense vegetation can help to distribute flows evenly across swales and slow velocities. Milder-sloped swales may have issues with water-logging and stagnant ponding, in which case underdrains can be employed to alleviate problems. If the swale is to be vegetated, vegetation must be capable of withstanding design flows and be of sufficient density to provide good filtration). Ideally, vegetation height should be above treatment flow water levels. If runoff enters directly into a swale, perpendicular to the main flow direction, the edge of the swale acts as a buffer and provides pre-treatment for the water entering the swale.

Ponds and Lakes

Ponds and Lakes are artificial bodies of open water that are usually created by constructing a dam wall with a weir outlet structure. Similar to constructed wetlands, they can be used to treat runoff by providing extended detention and allowing sedimentation, absorption of nutrients and UV disinfection to occur. In addition, they provide an aesthetic quality for recreation, wildlife habitat, and valuable storage of water that can potentially be reused for e.g. irrigation. Often, artificial ponds and lakes also form part of a flood detention system. Aquatic vegetation plays an important role for the water quality in artificial lakes and ponds in respect of maintaining and regulating the oxygen and nutrient levels. Due to a water depth greater than 1.5m, emergent macrophytes are usually restricted to the margins but submergent plants may occur in the open water zone. Fringing vegetation can be useful in reducing bank erosion. Ponds are normally not used as stand-alone WSUD measure but are often combined with sediment basins or constructed wetlands as pretreatments.

In many cases however, lakes and ponds have been designed as aesthetic features but suffer from poor health which can be caused by lack of appropriate inflows sustaining lake water levels, poor water quality of inflows and high organic carbon loads, infrequent flushing of the lake (too long residence time), and/or inappropriate mixing (stratification) leading to low levels of dissolved oxygen. Bluegreen algae caused by poor water quality and high nutrient levels can be a major threat to the health of lakes. To ensure the long-term sustainability of lakes and ponds, key issues that should be considered in their design include catchment hydrology and water level, and layout of the pond/lake (oriented to dominant winds to facilitate mixing. Hydraulic structures (inlet and outlet zones) should be designed to ensure adequate pre-treatment and prevent large nutrient 'spikes' Landscape design, using appropriate plant species and planting density are also necessary. High costs of the planned pond/lake area and of vegetation establishment as well as frequent maintenance requirements can be deterrents to use of ponds and lakes as WSUD measures.

The maintenance of pond and lake systems is important to minimise the risk of poor health. The inlet zone usually requires weed, plant, debris and litter removal with occasional replanting. In some cases, an artificial turn over of the lake might be necessary.

Water Re-use

Rainwater Tanks

Rainwater tanks are designed to conserve potable water by harvesting rain and stormwater to partially meet domestic water demands (e.g. during drought periods). In addition, rainwater tanks can reduce stormwater runoff volumes and stormwater pollutants from reaching downstream waterways. They can be used effectively in domestic households as a potential WSUD element. Rain and stormwater from rooftops of buildings can be collected and accessed specifically for purposes such as toilet flushing, laundry, garden watering and car washing. Buffer Tanks allow rain water collected from hard surfaces to seep into the site helps maintain the aquifer and ground water levels.

Rainwater tanks can reduce stormwater runoff volumes and stormwater pollutants from reaching downstream waterways. They can be used effectively in domestic households as a potential WSUD element. Rain and stormwater from rooftops of buildings can be collected and accessed specifically for purposes such as toilet flushing, laundry, garden watering and car washing.

In Australia, there are no quantitative performance targets for rainwater tanks, such as on size of tank or targeted reductions in potable water demand, in policies or guidelines. The various guidelines provided by state governments however, do advise that rain water tanks be designed to provide a reliable source of water to supplement mains water supply, and maintain appropriate water quality. The use of rainwater tanks should consider issues such as supply and demand, water quality, stormwater benefits (volume is reduced), cost, available space, maintenance, size, shape and material of the tank. Rainwater tanks must also be installed in accordance with plumbing and drainage standards. An advised suitable configuration may include a water filter or first flush diversion, a mains water top-up supply (dual supply system), maintenance drain, a pump (pressure system), and an on-site retention provision.

Potential water quality issues include atmospheric pollution, bird and possum droppings, insects e.g. mosquitoe larvae, roofing material, paints and detergents. As part of maintenance, an annual flush out (to remove built up sludge and debris) and regular visual inspections should be carried out.

Aquifer Storage and Recovery (ASR)

Aquifer storage and recovery (ASR) (also referred to as Managed Aquifer Recharge) aims to enhance water recharge to underground aquifers through gravity feed or pumping. It can be an alternative to large surface storages with water being pumped up again from below the surface in dry periods. Potential water sources for an ASR system can be stormwater or treated wastewater. The following components can usually be found in an ASR system that harvests stormwater:

1. A diversion structure for a stream or drain;

2. A treatment system for storm water prior to injection as well as for recovered water;

3. A wetland, detention pond, dam or tank, as a temporary storage measure;

4. A spill or overflow structure;

5. A well for the water injection and a well for the recovery of the water, and

6. Systems (including sampling ports) to monitor water levels and water quality.

The possible aquifer types suitable for an ASR system include fractured unconfined rock and confined sand and gravel. Detailed geological investigations are necessary to establish the feasibility of an ASR scheme. The potential low cost of ASR compared to subsurface storage can be attractive. The design process should consider the protection of groundwater quality, and recovered water quality for its intended use. Aquifers and aquitards need also be protected from damaged by depletion or high pressures. Impacts of the harvesting point on downstream areas also require consideration. Careful planning is required regarding aquifer selection, treatment, injection, the recovery process, and maintenance and monitoring.

Policy, Planning and Legislation

In Australia, due to the constitutional division of power between the Australian Commonwealth and the States, there is no national legislative requirement for urban water cycle management. The National Water Initiative (NWI), agreed upon by federal, state and territory governments in 2004 and 2006, provides a national plan to improve water management across the country. It provides clear intent to "Create Water Sensitive Australian Cities" and encourages adoption of WSUD approaches. National guidelines have also been released in accordance with NWI clause 92(ii) to provide guidance on evaluation of WSUD initiatives.

At the state level, planning and environmental legislation broadly promotes ecologically sustainable development, but to varying degrees have only limited requirements for WSUD. State planning policies variously provide more specific standards for adoption of WSUD practices in particular circumstances.

At the local government level, regional water resource management strategies supported by regional and/or local catchment-scale integrated water cycle management plans and/or stormwater management plans provide the strategic context for WSUD. Local government environment plans may place regulatory requirements on developments to implement WSUD.

As regulatory authority over stormwater runoff is shared between Australian states and local government areas, issues of multiple governing jurisdictions have resulted in inconsistent implementation of WSUD policies and practices and fragmented management of larger watersheds. For example, in Melbourne, jurisdictional authority for watersheds of greater than 60 ha rests with the state-level authority, Melbourne Water; while local governments govern smaller watersheds. Consequently, Melbourne Water has been deterred from investing significantly in WSUD works to improve small watersheds, despite them affecting the condition of the larger watersheds into which they drain.

State Legislation and Policy

Victoria

In Victoria, elements of WSUD are integrated into many of the overall objectives and strategies of the Victorian planning policyThe State Planning Policy Framework of the [Victoria Planning Provisions] which is contained in all planning schemes in Victoria contains some specific clauses requiring adoption of WSUD practices.

New residential developments are subject to a permeability standard that at least 20 per cent of sites should not be covered by impervious surfaces. The objective of this is to reduce the impact of increased stormwater run-off on the drainage system and facilitate on-site storm-water infiltration.

New residential subdivisions of two or more lots are required to meet integrated water management objectives related to:

- drinking water supply;

- reused and recycled water;

- waste water management, and

- urban run-off management.

Specifically regarding urban run-off management, the *Victoria Planning Provisions* c. 56.07-4 Clause 25 states that stormwater systems must meet best practice stormwater management objectives contained in the state guide Urban Stormwater: Best Practice Environmental Management Guidelines. The current water quality objectives are:

- 80 per cent retention of typical urban annual suspended solids load;

- 45 per cent retention of typical urban annual total phosphorus load;

- 45 per cent retention of typical urban annual total nitrogen load; and

- 70 per cent reduction if typical urban annual litter load.

Urban stormwater management systems must also meet the requirements of the relevant drainage authority. This is usually the local council. However, in the Melbourne region, where a catchment greater than 60ha is concerned it is Melbourne Water. Inflows downstream of the subdivision site are also restricted to pre-development levels unless approved by the relevant drainage authority and there are no detrimental downstream impacts.

Melbourne Water provides a simplified online software tool, STORM (Stormwater Treatment Objective – Relative Measure), to allow users to assess if development proposals meet legislated best practice stormwater quality performance objectives. The STORM tool is limited to assessment of discrete WSUD treatment practices and so does not model where several treatment practices are used in series. It is also limited to sites where coverage of impervious surfaces is greater than 40%. For larger more complicated developments more sophisticated modelling, such as MUSIC software, is recommended.

New South Wales

At the state level in New South Wales, the *State Environmental Planning Policy (Building Sustainability Index: BASIX) 2004* (NSW) is the primary piece of policy mandating adoption of WSUD. BASIX is an online program that allows users to enter data relating to a residential development, such as location, size, building materials etc.; to receive scores against water and energy use reduction targets. Water targets range from a 0 to 40% reduction in consumption of mains-supplied potable water, depending on location of the residential development. Ninety per cent of new homes are covered by the 40% water target. The BASIX program allows for the modelling of some WSUD elements such as use of rainwater tanks, stormwater tanks and greywater recycling.

Local Councils are responsible for the development of Local Environment Plans (LEPs) which can control development and mandate adoption of WSUD practices and targets *Local Government Act 1993* (NSW). Due to a lack of consistent policy and direction at the state-level however, adoption by local councils is mixed with some developing their own WSUD objectives in their local environmental plans (LEP) and others having no such provisions.

In 2006 the then NSW Department of Environment and Conservation released a guidance document, Managing Urban Stormwater: Harvesting and Reuse. The document presented an overview of stormwater harvesting and provided guidance on planning and design aspects of integrated landscape-scale strategy as well as technical WSUD practice implementation. The document now however, although still available on the governmental website, does not appear to be widely promoted.

The Sydney Metropolitan Catchment Management Authority also provides tools and resources to support local council adoption of WSUD. These include

- Potential WSUD provisions for incorporation into Local Government LEPs, with State-level department approval in NSW;

- Potential WSUD clauses for incorporation into Local Government reports, tenders, expressions of interest or other materials.;

- A WSUD Decision Support Tool to guide councils in comparing and evaluating on-ground WSUD projects, and

- Draft guidelines for the use of the more sophisticated MUSIC modelling software in NSW

Predictive Modelling to Assess WSUD Performance

Simplified modelling programs are provided by some jurisdictions to assess implementation of WSUD practices in compliance with local regulations. STORM is provided by Melbourne and BASIX is used in NSW for residential developments. For large, more complicated developments, more sophisticated modelling software may be necessary.

Issues Affecting Decision-making in WSUD

Impediments to the Adoption of WSUD

Major issues affecting the adoption of WSUD include:

- Regulatory framework barriers and institutional fragmentation at state and local government levels;

- Assessment and costing uncertainties relating to selecting and optimising WSUD practices for quantity and quality control;

- Technology and design and complexity integrating into landscape-scale water management systems; and

- Marketing and acceptance and related uncertainties.

The transition of Melbourne city to WSUD over the last four decades has culminated in a list of best practice qualities and enabling factors, which have been identified as important in aiding decision making to facilitate transition to WSUD technologies. The implementation of WSUD can be enabled through the effective interplay between the two variables discussed below.

Qualities of Decision-makers

- Vision for waterway health – A common vision for waterway health through cooperative approaches;

- Multi-sectoral network – A network of champions interacting across government, academia and private sector;

- Environmental values – Strong environmental protection values;

- Public-good disposition – Advocacy and protection of the public good;

- Best-practice ideology – Pragmatic approach to aid cross-sectoral implementation of best practices;

- Learning-by-doing philosophy – Adaptive approach to incorporating new scientific information;

- Opportunistic – Strategic and forward thinking approach to advocacy and practice, and

- Innovative and adaptive – Challenge status quo through focus on adaptive management philosophy.

Key Factors for Enabling WSUD

- Socio-political capital – An aligned community, media and political concern for improved waterway health, amenity and recreation;

- Bridging organisation – Dedicated organising entity that facilitates collaboration across science and policy, agencies and professions, and knowledge brokers and industry;

- Trusted and reliable science – Accessible scientific expertise, innovating reliable and effective solutions to local problems;

- Binding targets – A measurable and effective target that binds the change activity of scientists, policy makers and developers;

- Accountability – A formal organisational responsibility for the improvement of waterway health, and a cultural commitment to proactively influence practices that lead to such an outcome;

- Strategic funding – Additional resources, including external funding injection points, directed to the change effort;

- Demonstration projects and training – Accessible and reliable demonstration of new thinking and technologies in practice, accompanied by knowledge diffusion initiatives, and

- Market receptivity – A well-articulated business case for the change activity.

WSUD Projects in Australia

WSUD technologies can be implemented in a range of projects, from previously pristine and undeveloped, or Greenfield sites, to developed or polluted Brownfield sites that require alteration or remediation. In Australia, WSUD technologies have been implemented in a broad range of projects, including from small-scale roadside projects, up to large-scale +100 hectare residential development sites. The three key case studies below represent a range of WSUD projects from around Australia.

A Raingarden Biofilter for Small-scale Stormwater Management

Ku-ring-gai Council's Kooloona Crescent Raingarden, NSW

The WSUD Roadway Retrofit Bioretention System is a small-scale project implemented by the Ku-ring-gai Council in NSW as part of an overall catchment incentive to reduce stormwater pollution. The Raingarden uses a bioretention system to capture and treat an estimated 75 kg of total suspended solids (TSS) per year of local stormwater runoff from the road, and filters it through a sand filter media before releasing it back into the stormwater system. Permeable pavers are also used in the system within the surrounding pedestrian footpaths, to support the infiltration of runoff into the ground water system. Roadside bioretention systems similar to this project have been implemented throughout Australia. Similar projects are presented on the Sydney Catchment Management Authority's WSUD website:

- *2005 Ku-ring-gai Council* – Minnamurra Avenue Water Sensitive Road Retrofit Project;

- *2003 City of Yarra, Victoria* – Roadway reconstruction with inclusion of bio-retention basins to treat stormwater;

- *2003-4 City of Kingston, Victoria (Chelsea)* – Roadway reconstruction with inclusion of bioretention basins to treat stormwater, and

- *2004 City of Kingston, Victoria (Mentone)* – Roadway reconstruction with inclusion of bioretention basins to treat stormwater.

WSUD in Residential Development Projects

Lynbrook Estate, Victoria

The Lynbrook Estate development project in Victoria, demonstrates effective implementation of WSUD by the private sector. It is a Greenfield residential development site that has focused its marketing for potential residents on innovative use of stormwater management technologies, following a pilot study by Melbourne Water.

The project combines conventional drainage systems with WSUD measures at the streetscape and sub-catchment level, with the aim of attenuating and treating stormwater flows to protect receiving waters within the development. Primary treatment of the stormwater is carried out by grass swales and an underground gravel trench system, which collects, infiltrates and conveys road/roof runoff . The main boulevard acts as a bioretention system with an underground gravel filled trench to allow for infiltration and conveyance of stormwater. The catchment runoff then undergoes secondary treatment through a wetland system before discharge into an ornamental lake. This project is significant as the first residential WSUD development of this scale in Australia. Its performance in exceeding the Urban Stormwater Best Practice Management Guidelines for Total Nitrogen, Total Phosphorus and Total Suspended Solids levels, has won it both the 2000 President's Award in the Urban Development Institute of Australia Awards for Excellence (recognising innovation in urban development), and the 2001 Cooperative Research Centres' Association Technology Transfer Award. Its success as a private-sector implemented WSUD system led to its proponent Urban and Regional Land Corporation (URLC) to look to incorporate WSUD as a standard practice across the State of Victoria. The project has also attracted attention from developers, councils, waterway management agencies and environmental policy-makers throughout the country.

Large-scale Remediation for the Sydney 2000 Olympic Games

Homebush Bay, NSW

For the establishment of the Sydney 2000 Olympic Games site, the Brownfield area of Homebush Bay was remediated from an area of landfill, abattoirs and a navy armament depots into a multiuse Olympic site. A Water Reclamation and Management Scheme

(WRAMS) was set up in 2000 for large-scale recycling of non-potable water, which included a range of WSUD technologies. These technologies were implemented with a particular focus on addressing the objectives of protecting receiving waters from stormwater and wastewater discharges; minimising potable water demand; and protecting and enhancing habitat for threatened species 2006. The focus of WSUD technologies was directed towards the on-site treatment, storage and recycling of stormwater and wastewater. Stormwater runoff is treated using gross pollutant traps, swales and/or wetland systems. This has contributed to a reduction of 90% in nutrient loads in the Haslams Creek wetland remediation area. Wastewater is treated in a water reclamation plant. Almost 100% of sewage is treated and recycled. The treated water from both stormwater and wastewater sources is stored and recycled for use throughout the Olympic site in water features, irrigation, toilet flushing and fire fighting capacities. Through the use of WSUD technology, the WRAMS scheme has resulted in the conservation of 850 million litres (ML) of water annually, a potential 50% reduction in annual potable water consumption within the Olympic site, as well as the annual diversion of approximately 550 ML of sewage normally discharged through ocean outfalls. As part of the long-term sustainability focus of the 'Sydney Olympic Park Master Plan 2030', the Sydney Olympic Park Authority (SOPA) has identified key best practice environmental sustainability approaches to include, the connection to recycled water and effective water demand management practices, maintenance and extension of recycled water systems to new streets as required, and maintenance and extension of the existing stormwater system that recycles water, promotes infiltration to sub soil, filters pollutants and sediments, and minimises loads on adjoining waterways. The SOPA has used WSUD technology to ensure that the town remains 'nationally and internationally recognised for excellence and innovation in urban design, building design and sustainability, both in the present and for future generations.

References

- Linsley, Ray K. & Franzini, Joseph B. Water-Resources Engineering (1972) McGraw-Hill ISBN 0-07-037959-9 pp.454-456

- Franson, Mary Ann (1975). Standard Methods for the Examination of Water and Wastewater 14th ed. Washington, DC: American Public Health Association, American Water Works Association & Water Pollution Control Federation. ISBN 0-87553-078-8

- Franson, Mary Ann Standard Methods for the Examination of Water and Wastewater 14th edition (1975) APHA, AWWA & WPCF ISBN 0-87553-078-8 pp.125-126

- Air pollution, acid rain, and the environment. Kenneth Mellanby, Watt Committee on Energy, Springer, 1988 ISBN 1-85166-222-7, ISBN 978-1-85166-222-7

- "Center for Coastal Monitoring and Assessment: Mussel Watch Contaminant Monitoring". Ccma. nos.noaa.gov. 2014-01-14. Retrieved 2015-09-04.

- International Organization for Standardization (ISO). "13.060: Water quality". Geneva, Switzerland. Retrieved 2011-07-04.

Radiation Monitoring: An Integrated Study

Radiation monitoring involves the measurement of radiation dose. It indicates the toxicity of radioactive substances. Some of the topics discussed in this chapter are absorbed dose, radionuclide, ionization chamber, gaseous ionization detectors and scintillation counter. This chapter will provide an integrated understanding of radiation monitoring.

Radiation Monitoring

Radiation monitoring involves the measurement of radiation dose or radionuclide contamination for reasons related to the assessment or control of exposure to radiation or radioactive substances, and the interpretation of the results.

The U.S. Navy monitored radiation from the Fukushima I nuclear accidents

Environmental Monitoring

Environmental monitoring is the measurement of external dose rates due to sources in the environment or of radionuclide concentrations in environmental media.

Source Monitoring

Source monitoring is a specific term used in ionising radiation monitoring, and according to the IAEA, is the measurement of activity in radioactive material being released to the environment or of external dose rates due to sources within a facility or activity.

In this context a source is anything that may cause radiation exposure — such as by emitting ionising radiation, or releasing radioactive substances. The phrase "standard source" is also used as a de facto term in the more specific context of being a calibration standard source in ionising radiation metrology.

Personnel wearing full protective gear while checking air filters for radioactive contamination

The methodological and technical details of the design and operation of source and environmental radiation monitoring programmes and systems for different radionuclides, environmental media and types of facility are given in IAEA Safety Standards Series No. RS–G-1.8 and in IAEA Safety Reports Series No. 64.

Radiation Protection Instruments

Practical radiation measurement using calibrated radiation protection instruments is essential in evaluating the effectiveness of protection measures, and in assessing the radiation dose likely to be received by individuals. The measuring instruments for radiation protection are both "installed" (in a fixed position) and portable (hand-held or transportable).

Installed Instruments

Installed instruments are fixed in positions which are known to be important in assessing the general radiation hazard in an area. Examples are installed "area" radiation monitors, Gamma interlock monitors, personnel exit monitors, and airborne particulate monitors.

The area radiation monitor will measure the ambient radiation, usually X-Ray, Gamma or neutrons; these are radiations which can have significant radiation levels over a range in excess of tens of metres from their source, and thereby cover a wide area.

Gamma radiation "interlock monitors" are used in applications to prevent inadvertent exposure of workers to an excess dose by preventing personnel access to an area when a high radiation level is present. These interlock the process access directly.

Airborne contamination monitors measure the concentration of radioactive particles in the ambient air to guard against radioactive particles being ingested, or deposited in the lungs of personnel. These instruments will normally give a local alarm, but are often connected to an integrated safety system so that areas of plant can be evacuated and personnel are prevented from entering an air of high airborne contamination.

Personnel exit monitors (PEM) are used to monitor workers who are exiting a "contamination controlled" or potentially contaminated area. These can be in the form of hand monitors, clothing frisk probes, or whole body monitors. These monitor the surface of the workers body and clothing to check if any radioactive contamination has been deposited. These generally measure alpha or beta or gamma, or combinations of these.

The UK National Physical Laboratory publishes a good practice guide through its Ionising Radiation Metrology Forum concerning the provision of such equipment and the methodology of calculating the alarm levels to be used.

Portable Instruments

Portable instruments are hand-held or transportable. The hand-held instrument is generally used as a survey meter to check an object or person in detail, or assess an area where no installed instrumentation exists. They can also be used for personnel exit monitoring or personnel contamination checks in the field. These generally measure alpha, beta or gamma, or combinations of these.

Hand-held ion chamber survey meter in use

Transportable instruments are generally instruments that would have been permanently installed, but are temporarily placed in an area to provide continuous monitoring where it is likely there will be a hazard. Such instruments are often installed on trolleys to allow easy deployment, and are associated with temporary operational situations.

In the United Kingdom the HSE has issued a user guidance note on selecting the correct radiation measurement instrument for the application concerned. This covers all radiation instrument technologies, and is a useful comparative guide.

Instrument Types

A number of commonly used detection instruments are listed below.

- ionization chambers
- proportional counters
- Geiger counters
- Semiconductor detectors
- Scintillation detectors
- Airborne particulate radioactivity monitoring

The links should be followed for a fuller description of each.

Absorbed Dose

Absorbed dose is a physical dose quantity D representing the mean energy imparted to matter per unit mass by ionizing radiation. In the SI system of units, the unit of measure is joules per kilogram, and its special name is gray (Gy). The non-SI CGS unit rad is sometimes also used, predominantly in the USA.

Uses

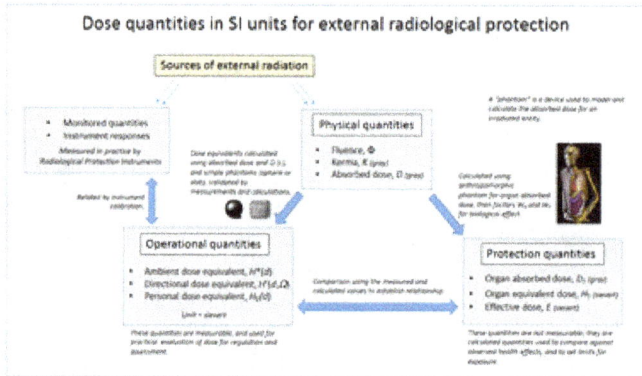

External dose quantities used in radiation protection and dosimetry

Absorbed dose is used in the calculation of dose uptake in living tissue in both radiation protection and radiology. It is also used to directly compare the effect of radiation on inanimate matter.

Radiological Protection

The quantity absorbed dose is of fundamental importance in radiological protection for calculating radiation dose. However, absorbed dose is a physical quantity and used unmodified is not an adequate indicator of the likely health effects in humans.

It has been found that for stochastic radiation risk (defined as *probability* of cancer induction and genetic effects) consideration must be given to the type of radiation and the sensitivity of the irradiated tissues, which requires the use of modifying factors. Conventionally therefore, unmodified absorbed dose is not used for comparing stochastic risks but only used to compare against deterministic effects (*severity* of acute tissue effects that are certain to happen) such as in acute radiation syndrome.

To represent stochastic risk the equivalent dose H_T and effective dose E are used, and appropriate dose factors and coefficients are used to calculate these from the absorbed dose. Equivalent and effective dose quantities are expressed in units of the sievert or rem which implies that biological effects have been taken into account. These are usually in accordance with the recommendations of the International Committee on

Radiation Protection (ICRP) and International Commission on Radiation Units and Measurements (ICRU). The coherent system of radiological protection quantities developed by them is shown in the accompanying diagram.

Radiology

The measurement of absorbed dose in tissue is of fundamental importance in radiobiology and radiation therapy as it is the measure of the amount of energy the incident radiation is imparting to the target tissue.

Component Survivability

Absorbed dose is used to rate the survivability of devices such as electronic components in ionizing radiation environments.

Food Irradiation

The international Radura logo, used to show a food has been treated with ionizing radiation.

Absorbed dose is the physical dose quantity used to ensure irradiated food has received the correct dose to ensure effectiveness. Variable doses are used depending on the application and can be as high as 70 kGy.

Computation

The absorbed dose is equal to the radiation exposure (ions or C/kg) of the radiation beam multiplied by the ionization energy of the medium to be ionized.

For example, the ionization energy of dry air at 20 °C and 101.325 kPa of pressure is 33.97±0.06 J/C. (33.97 eV per ion pair) Therefore an exposure of 2.58×10^{-4} C/kg (1 roentgen) would deposit an absorbed dose of 8.76×10^{-3} J/kg (0.00876 Gy or 0.876 rad) in dry air at those conditions.

When the absorbed dose is not uniform, or when it is only applied to a portion of a body or object, an absorbed dose representative of the entire item can be calculated by taking a mass-weighted average of the absorbed doses at each point.

More precisely,

$$\overline{D_T} = \frac{\int_T D(x,y,z)\rho(x,y,z)dV}{\int_T \rho(x,y,z)dV}$$

Where

$\overline{D_T}$ is the mass-averaged absorbed dose of the entire item T

T is the item of interest

$D(x,y,z)$ is the absorbed dose as a function of location

$\rho(x,y,z)$ is the density as a function of location

V is volume

Medical Considerations

Non-uniform absorbed dose is common for soft radiations such as low energy x-rays or beta radiation. Self-shielding means that the absorbed dose will be higher in the tissues facing the source than deeper in the body.

The mass average can be important in evaluating the risks of radiotherapy treatments, since they are designed to target very specific volumes in the body, typically a tumour. For example, if 10% of a patient's bone marrow mass is irradiated with 10 Gy of radiation locally, then the absorbed dose in bone marrow overall would be 1 Gy. Bone marrow makes up 4% of the body mass, so the whole-body absorbed dose would be 0.04 Gy. The first figure (10 Gy) is indicative of the local effects on the tumour, while the second and third figure (1 Gy and 0.04 Gy) are better indicators of the overall health effects on the whole organism. Additional dosimetry calculations would have to be performed on these figures to arrive at a meaningful effective dose, which is needed to estimate the risk of cancer or other stochastic effects.

When ionizing radiation is used to treat cancer, the doctor will usually prescribe the radiotherapy treatment in units of gray. Medical imaging doses may be described in units of coulomb per kilogram, but when radiopharmaceuticals are used, they will usually be administered in units of becquerel.

Radiation-related Quantities

The following table shows radiation quantities in SI and non-SI units.

Quantity	Name	Symbol	Unit	Year	System
Exposure (X)	röntgen	R	esu / 0.001293 g of air	1928	non-SI

Absorbed dose (D)			erg•g⁻¹	1950	non-SI
	rad	rad	100 erg•g⁻¹	1953	non-SI
	gray	Gy	J•kg⁻¹	1974	SI
Activity (A)	curie	Ci	3.7×10^{10} s⁻¹	1953	non-SI
	becquerel	Bq	s⁻¹	1974	SI
Dose equivalent (H)	röntgen equivalent man	rem	100 erg•g⁻¹	1971	non-SI
	sievert	Sv	J•kg⁻¹	1977	SI
Fluence (Φ)	(reciprocal area)		cm⁻² or m⁻²	1962	SI (m⁻²)

Although the United States Nuclear Regulatory Commission permits the use of the units curie, rad, and rem alongside SI units, the European Union European units of measurement directives required that their use for "public health ... purposes" be phased out by 31 December 1985.

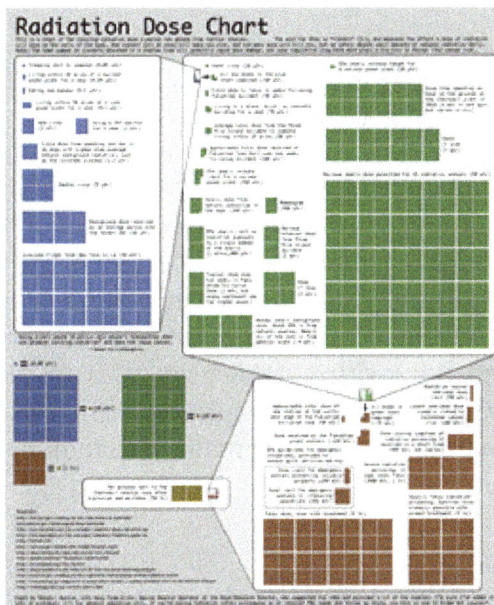

Various doses of radiation in sieverts, ranging from trivial to lethal.

Radionuclide

A radionuclide (radioactive nuclide, radioisotope or radioactive isotope) is an atom that has excess nuclear energy, making it unstable. This excess energy can either create and emit, from the nucleus, new radiation (gamma radiation) or a new particle (alpha particle or beta particle), or transfer this excess energy to one of its electrons, causing it to be ejected (conversion electron). During this process, the radionuclide is said to undergo radioactive decay. These emissions constitute ionizing radiation. The unstable nucleus is more stable following the emission, but will sometimes undergo further decay. Radioactive decay is a random process at the level of single atoms: it is impossible to predict when one particular atom will decay. However, for a collection of atoms of a single element the decay rate, and thus the half-life ($t_{1/2}$) for that collection can be calculated from their measured decay constants. The duration of the half-lives of radioactive atoms have no known limits; the time range is over 55 orders of magnitude.

Radionuclides both occur naturally and are artificially made using nuclear reactors, cyclotrons, particle accelerators or radionuclide generators. There are about 650 radio-nuclides with half-lives longer than 60 minutes. Of these, 34 are primordial radionuclides that existed before the creation of the solar system, and there are another 50 radionuclides detectable in nature as daughters of these, or produced naturally on Earth by cosmic radiation. More than 2400 radionuclides have half-lives less than 60 minutes. Most of these are only produced artificially, and have very short half-lives. For comparison, there are about 254 stable nuclides.

All chemical elements have radionuclides. Even the lightest element, hydrogen, has a well-known radionuclide, tritium. Elements heavier than lead, and the elements technetium and promethium, exist only as radionuclides.

Unplanned exposure to radionuclides generally has a harmful effect on living organisms including humans, although low levels of exposure occur naturally without harm. The degree of harm will depend on the nature and extent of the radiation produced, the amount and nature of exposure (close contact, inhalation or ingestion), and the biochemical properties of the element; with increased risk of cancer the most usual consequence. However, radionuclides with suitable properties are used in nuclear medicine for both diagnosis and treatment. An imaging tracer made with radionuclides is called a radioactive tracer. A pharmaceutical drug made with radionuclides is called a radiopharmaceutical.

Origin

Natural

On Earth, naturally occurring radionuclides fall into three categories: primordial radionuclides, secondary radionuclides, and cosmogenic radionuclides.

- Radionuclides are produced in stellar nucleosynthesis and supernova explosions along with stable nuclides. Most decay quickly but can still be observed astronomically and can play a part in understanding astronomic processes. Primordial radionuclides, such as uranium and thorium, exist in the present time because their half-lives are so long (>80 million years) they have not yet completely decayed. Some radionuclides have half-lives so long (many times the age of the universe) that decay has only recently been detected, and for most practical purposes they can be considered stable, most notably bismuth-209:

detection of this decay meant that bismuth was no longer considered stable. It is possible decay may be observed in other nuclides adding to this list of primordial radionuclides.

- Secondary radionuclides are radiogenic isotopes derived from the decay of primordial radionuclides. They have shorter half-lives than primordial radionuclides. They arise in the decay chain of the primordial isotopes thorium-232, uranium-238 and uranium-235. Examples include the natural isotopes of polonium and radium.

- Cosmogenic isotopes, such as carbon-14, are present because they are continually being formed in the atmosphere due to cosmic rays.

Many of these radionuclides exist only in trace amounts in nature, including the two shortest-lived primordial nuclides and all cosmogenic nuclides. Secondary radionuclides will occur in proportion to their half-lives, so short-lived ones will be very rare. Thus polonium can be found in uranium ores at about 0.1 mg per metric ton (1 part in 10^{10}),. Further radionunclides may occur in nature in virtually undetectable amounts as a result of rare events such as spontaneous fission or uncommon cosmic ray interactions.

Nuclear Fission

Radionuclides are produced as an unavoidable result of nuclear fission and thermonuclear explosions. The process of nuclear fission creates a wide range of fission products, most of which are radionuclides. Further radionuclides can be created from irradiation of the nuclear fuel (creating a range of actinides) and of the surrounding structures, yielding activation products. This complex mixture of radionuclides with different chemistries and radioactivity makes handling nuclear waste and dealing with nuclear fallout particularly problematic.

Synthetic

Synthetic radionuclides are deliberately synthesised using nuclear reactors, particle accelerators or radionuclide generators:

- As well as being extracted from nuclear waste, radioisotopes can be produced deliberately with nuclear reactors, exploiting the high flux of neutrons present. These neutrons activate elements placed within the reactor. A typical product from a nuclear reactor is iridium-192. The elements that have a large propensity to take up the neutrons in the reactor are said to have a high neutron cross-section.

- Particle accelerators such as cyclotrons accelerate particles to bombard a target to produce radionuclides. Cyclotrons accelerate protons at a target to produce positron-emitting radionuclides, e.g. fluorine-18.

- Radionuclide generators contain a parent radionuclide that decays to produce a radioactive daughter. The parent is usually produced in a nuclear reactor. A typical example is the technetium-99m generator used in nuclear medicine. The parent produced in the reactor is molybdenum-99.

Uses

Radionuclides are used in two major ways: either for their radiation alone (irradiation, nuclear batteries) or for the combination of chemical properties and their radiation (tracers, biopharmaceuticals).

- In biology, radionuclides of carbon can serve as radioactive tracers because they are chemically very similar to the nonradioactive nuclides, so most chemical, biological, and ecological processes treat them in a nearly identical way. One can then examine the result with a radiation detector, such as a Geiger counter, to determine where the provided atoms were incorporated. For example, one might culture plants in an environment in which the carbon dioxide contained radioactive carbon; then the parts of the plant that incorporate atmospheric carbon would be radioactive. Radionuclides can be used to monitor processes such as DNA replication or amino acid transport.

- In nuclear medicine, radioisotopes are used for diagnosis, treatment, and research. Radioactive chemical tracers emitting gamma rays or positrons can provide diagnostic information about internal anatomy and the functioning of specific organs, including the human brain. This is used in some forms of tomography: single-photon emission computed tomography and positron emission tomography (PET) scanning and Cherenkov luminescence imaging. Radioisotopes are also a method of treatment in hemopoietic forms of tumors; the success for treatment of solid tumors has been limited. More powerful gamma sources sterilise syringes and other medical equipment.

- In food preservation, radiation is used to stop the sprouting of root crops after harvesting, to kill parasites and pests, and to control the ripening of stored fruit and vegetables.

- In industry, and in mining, radionuclides are used to examine welds, to detect leaks, to study the rate of wear, erosion and corrosion of metals, and for on-stream analysis of a wide range of minerals and fuels.

- In spacecraft and elsewhere, radionuclides are used to provide power and heat, notably through radioisotope thermoelectric generators (RTGs).

- In astronomy and cosmology radionuclides play a role in understanding stellar and planetary process.

- In particle physics, radionuclides help discover new physics (physics beyond the Standard Model) by measuring the energy and momentum of their beta decay products.

- In ecology, radionuclides are used to trace and analyze pollutants, to study the movement of surface water, and to measure water runoffs from rain and snow, as well as the flow rates of streams and rivers.

- In geology, archaeology, and paleontology, natural radionuclides are used to measure ages of rocks, minerals, and fossil materials.

Examples

The following table lists properties of selected radionuclides illustrating the range of properties and uses.

Key: Z = no of protons; N = no of Neutrons; DM = Decay Mode; DE = Decay Energy

Americium-241

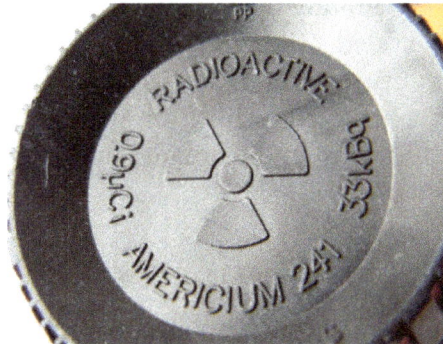

Americium-241 container in a smoke detector.

Americium-241 capsule as found in smoke detector. The circle of darker metal in the center is americium-241; the surrounding casing is aluminium.

Most household smoke detectors contain americium produced in nuclear reactors. The radioisotope used is americium-241. The element americium is created by bombarding plutonium with neutrons in a nuclear reactor. Its isotope americium-241 decays by emitting alpha particles and gamma radiation to become neptunium-237. Most common household smoke detectors use a very small quantity of ^{241}Am (about 0.29 micrograms per smoke detector) in the form of americium dioxide. Smoke detectors use ^{241}Am since the alpha particles it emits collide with oxygen and nitrogen particles in the air. This occurs in the detector's ionization chamber where it produces charged particles or ions. Then, these charged particles are collected by a small electric voltage that will create an electric current that will pass between two electrodes. Then, the ions that are flowing between the electrodes will be neutralized when coming in contact with smoke, thereby decreasing the electric current between the electrodes, which will activate the detector's alarm.

Gadolinium-153

The ^{153}Gd isotope is used in X-ray fluorescence and osteoporosis screening. It is a gamma-emitter with an 8-month half-life, making it easier to use for medical purposes. In nuclear medicine, it serves to calibrate the equipment needed like single-photon emission computed tomography systems (SPECT) to make x-rays. It ensures that the machines work correctly to produce images of radioisotope distribution inside the patient. This isotope is produced in a nuclear reactor from europium or enriched gadolinium. It can also detect the loss of calcium in the hip and back bones, allowing the ability to diagnose osteoporosis.

Impacts on Organisms

Radionuclides that find their way into the environment may cause harmful effects as radioactive contamination. They can also cause damage if they are excessively used during treatment or in other ways exposed to living beings, by radiation poisoning. Potential health damage from exposure to radionuclides depends on a number of factors, and "can damage the functions of healthy tissue/organs. Radiation exposure can produce effects ranging from skin redness and hair loss, to radiation burns and acute radiation syndrome. Prolonged exposure can lead to cells being damaged and in turn lead to cancer. Signs of cancerous cells might not show up until years, or even decades, after exposure."

Summary Table for Classes of Nuclides, "Stable" and Radioactive

Following is a summary table for the total list of nuclides with half-lives greater than one hour. Ninety of these 905 nuclides are theoretically stable, except to proton-decay (which has never been observed). About 254 nuclides have never been observed to decay, and are classically considered stable.

The remaining 650 radionuclides have half-lives longer than 1 hour, and are well-characterized. They include 28 nuclides with measured half-lives longer than the estimated age of the universe (13.8 billion years), and another 6 nuclides with half-lives long enough (> 80 million years) that they are radioactive primordial nuclides, and may be detected on Earth, having survived from their presence in interstellar dust since before the formation of the solar system, about 4.6 billion years ago. Another ~51 short-lived nuclides can be detected naturally as daughters of longer-lived nuclides or cosmic-ray products. The remaining known nuclides are known solely from artificial nuclear transmutation.

Numbers are not exact, and may change slightly in the future, as "stable nuclides" are observed to be radioactive with very long half-lives.

This is a summary table for the 905 nuclides with half-lives longer than one hour (including those that are stable), given in list of nuclides.

Stability class	Number of nuclides	Running total	Notes on running total
Theoretically stable to all but proton decay	90	90	Includes first 40 elements. Proton decay yet to be observed.
Energetically unstable to one or more known decay modes, but no decay yet seen. Spontaneous fission possible for "stable" nuclides ≥ niobium-93; other mechanisms possible for heavier nuclides. All considered "stable" until decay detected.	164	254	Total of classically stable nuclides.
Radioactive primordial nuclides.	34	288	Total primordial elements include uranium, thorium, bismuth, rubidium-87, potassium-40 plus all stable nuclides.
Radioactive nonprimordial, but naturally occurring on Earth.	~ 51	~ 339	Carbon-14 (and other isotopes generated by cosmic rays) and daughters of radioactive primordial elements, such as radium, polonium, etc.
Radioactive synthetic (half-life ≥ 1.0 hour). Includes most useful radiotracers.	556	905	These 905 nuclides are listed in the article List of nuclides.
Radioactive synthetic (half-life < 1.0 hour).	>2400	>3300	Includes all well-characterized synthetic nuclides.

List of Commercially Available Radionuclides

This list covers common isotopes, most of which are available in very small quantities to the general public in most countries. Others that are not publicly accessible are traded commercially in industrial, medical, and scientific fields and are subject to government regulation. For a complete list of all known isotopes for every element (minus activity data), see List of nuclides and Isotope lists. For a table, see Table of nuclides.

Gamma Emission Only

Isotope	Activity	Half-life	Energies (keV)
Barium-133	9694 TBq/kg (262 Ci/g)	10.7 years	81.0, 356.0
Cadmium-109	96200 TBq/kg (2600 Ci/g)	453 days	88.0
Cobalt-57	312280 TBq/kg (8440 Ci/g)	270 days	122.1
Cobalt-60	40700 TBq/kg (1100 Ci/g)	5.27 years	1173.2, 1332.5
Europium-152	6660 TBq/kg (180 Ci/g)	13.5 years	121.8, 344.3, 1408.0
Manganese-54	287120 TBq/kg (7760 Ci/g)	312 days	834.8
Sodium-22	237540 Tbq/kg (6240 Ci/g)	2.6 years	511.0, 1274.5
Zinc-65	304510 TBq/kg (8230 Ci/g)	244 days	511.0, 1115.5
Technetium-99m	$1.95{\times}10^4$ TBq/g (5.27×10^7 Ci/g)	6 hours	140

Beta Emission Only

Isotope	Activity	Half-life	Energies (keV)
Strontium-90	5180 TBq/kg (140 Ci/g)	28.5 years	546.0
Thallium-204	17057 TBq/kg (461 Ci/g)	3.78 years	763.4
Carbon-14	166.5 TBq/kg (4.5 Ci/g)	5730 years	49.5 (average)
Tritium (Hydrogen-3)	357050 TBq/kg (9650 Ci/g)	12.32 years	5.7 (average)

Alpha Emission Only

Isotope	Activity	Half-life	Energies (keV)
Polonium-210	166500 TBq/kg (4500 Ci/g)	138.376 days	5304.5
Uranium-238	12580 KBq/kg (0.00000034 Ci/g)	4.468 billion years	4267

Multiple Radiation Emitters

Isotope	Activity	Half-life	Radiation types	Energies (keV)
Caesium-137	3256 TBq/kg (88 Ci/g)	30.1 years	Gamma & beta	G: 32, 661.6 B: 511.6, 1173.2
Americium-241	129.5 TBq/kg (3.5 Ci/g)	432.2 years	Gamma & alpha	G: 59.5, 26.3, 13.9 A: 5485, 5443

Airborne Particulate Radioactivity Monitoring

Continuous particulate air monitors (CPAMs) have been used for years in nuclear facilities to assess airborne particulate radioactivity (APR). In more recent times they may also be used to monitor people in their homes for the presence of manmade radioactivity. These monitors can be used to trigger alarms, indicating to personnel that they should evacuate an area. This article will focus on CPAM use in nuclear power plants, as opposed to other nuclear fuel-cycle facilities, or laboratories, or public-safety applications.

In nuclear power plants, CPAMs are used for measuring releases of APR from the facility, monitoring levels of APR for protection of plant personnel, monitoring the air in the reactor containment structure to detect leakage from the reactor systems, and to control ventilation fans, when the APR level has exceeded a defined threshold in the ventilation system.

Introduction

CPAMs use a pump to draw air through a filter medium to collect airborne particulate matter that carries very small particles of radioactive material; the air itself is not radioactive. The particulate radioactive material might be natural, e.g., radon decay products ("progeny", e.g., ^{212}Pb), or manmade, usually fission or activation products (e.g., ^{137}Cs), or a combination of both. There are also "gas monitors" which pass the air through a sample chamber volume which is viewed continuously by a radiation detector. Radionuclides that occur in the gaseous form (e.g., ^{85}Kr) are not collected on the CPAM filter to any appreciable extent, so that a separate monitoring system is needed to assess these nuclide concentrations in the sampled air. These gas monitors are often placed downstream of a CPAM so that any particulate matter in the sampled air is collected by the CPAM and thus will not contaminate the gas monitor's sample chamber.

Monitoring vs. Sampling

In monitoring, the region of deposition of this material onto the filter medium is *continuously* viewed by a radiation detector, concurrent with the collection. This is as opposed to a sampling system, in which the airborne material is collected by pumping air, usually at a much higher volumetric flowrate than a CPAM, through a collection medium for some period of time, but there is no continuous radiation detection; the filter medium is removed *periodically* from the sampler and taken to a separate radiation detection system for analysis.

In general, sampling has better detection sensitivity for low levels of airborne radioactivity, due to the much larger total volume of air passing through the filter medium over the sampling interval (which may be on the order of hours), and also due to the more sophisticated forms of quantitative analysis available once the filter medium is removed from the sampler. On the other hand, monitoring with CPAMs provides near-ly real-time airborne radioactivity level indication. It is common practice to refer to "sampled" air even when discussing a CPAM, i.e., as opposed to "monitored" air, which would, strictly, be more correct.

CPAM Types

There are two major types of CPAMs, fixed-filter and moving-filter. In the former, the filter medium does not move while the airborne material is collected. The latter type has two main variants, the rectangular deposition area ("window") and the circular window. In both types of CPAM the sampled air is pulled (not pushed) by a pump through the piping of the monitor up to the structure that holds the filter medium. It is important to note that CPAM pumps are specially designed to maintain a constant volumetric flowrate.

As the air passes through the collection medium (usually a form of filter paper), par-ticulate matter is deposited onto the filter in either a rectangular or circular pattern, depending on the instrument's design, and then the air continues on its way out of the monitor. The *entire* deposition area, regardless of its geometric shape, is assumed to be viewed by a radiation detector of a type appropriate for the nuclide in question.

Moving-filter monitors are often used in applications where loading of the filter me-dium with dust is an issue; this dust loading reduces the air flow over time. The mov-ing-filter collection medium ("tape") is assumed to move across the deposition area at a constant, known rate. This rate is often established in such a way that a roll of the filter tape will last about one month; a typical filter movement rate is about one inch per hour.

The rectangular-window moving filter monitor will be denoted as RW, and the circular, CW. Fixed filter is FF.

CPAM Applications

Effluent Monitoring

CPAMs are used to monitor the air effluents from nuclear facilities, notably power reac-tors. Here the objective is to assess the amount of certain radionuclides released from the facility. Real-time measurement of the very low concentrations released by these facilities is difficult; a more-reliable measurement of the *total* radioactivity released over some time interval (days, perhaps weeks) may in some cases be an acceptable approach. In effluent monitoring, a sample of the air in the plant stack is withdrawn and pumped (pulled) down to the CPAM location. This sampled air in many cases must travel a considerable distance through piping. Extracting and transporting the particu-lates for the CPAM to measure in such a way that the measurement is representative of what is being released from the facility is challenging.

In the USA there are effluent monitoring requirements in both 10CFR20 and 10CFR50; Appendix B to the former and Appendix I to the latter are especially important. 10CFR50 Appendix A states:

> Criterion 64--Monitoring radioactivity releases. Means shall be provided for monitoring the reactor containment atmosphere, spaces containing components for recirculation of loss-of-coolant accident fluids, effluent discharge paths, and the plant environs for radioactivity that may be released from normal operations, including anticipated operational occurrences, and from postulated accidents.

Also in the USA, Regulatory Guide 1.21, *Measuring, Evaluating, and Reporting Radioactivity in Solid Wastes and Releases of Radioactive Materials in Liquid and Gaseous Effluents from Light-Water-Cooled Nuclear Power Plants* is highly relevant to this CPAM application.

Occupational Exposure Assessment

For occupational exposure (inhalation) assessment, CPAMs may be used to monitor the air in some volume, such as a compartment in a nuclear facility where personnel are working. A difficulty with this is that, unless the air in the compartment is uniformly mixed, the measurement made at the monitor location may not be representative of the concentration of radioactive material in the air that the workers are breathing. For this application the CPAM may be physically placed directly in the occupied compartment, or it may extract sampled air from the HVAC system that serves that compartment. The following portions of 10CFR20 are relevant to the requirement for occupational exposure CPAM applications in the USA: 10CFR20.1003 (definition of Airborne Radioactivity Area), 1201, 1204, 1501, 1502, 2103.

Process Monitoring and Control

Radiation monitors in general have a number of process-control applications in nuclear power plants; a major CPAM application in this area is the monitoring of the air intake for the plant control room. In the event of an accident, high levels of airborne radioactivity could be brought into the control room by its HVAC system; the CPAM monitors this air and is intended to detect high concentrations of radioactivity and shut down the HVAC flow when necessary.

For use in the USA, standard 10CFR50 Appendix A states:

> Criterion 19--Control room. A control room shall be provided from which actions can be taken to operate the nuclear power unit safely under normal conditions and to maintain it in a safe condition under accident conditions, including loss-of-coolant accidents. Adequate radiation protection shall be provided to permit access and occupancy of the control room under accident conditions

without personnel receiving radiation exposures in excess of 5 rem whole body, or its equivalent to any part of the body, for the duration of the accident. Equipment at appropriate locations outside the control room shall be provided (1) with a design capability for prompt hot shutdown of the reactor, including necessary instrumentation and controls to maintain the unit in a safe condition during hot shutdown, and (2) with a potential capability for subsequent cold shutdown of the reactor through the use of suitable procedures.

This defines a requirement for monitoring the air intake for the control room, such that the exposure limits, including for inhalation exposure, shall not be exceeded. CPAMs are often used for this.

Reactor Leak Detection

Leakage from the so-called "reactor coolant pressure boundary" is required to be monitored in USA nuclear power plants. Monitoring the airborne particulate radioactivity in the reactor containment structure is an acceptable method to meet this requirement, and so CPAMs are used. It is the case that when primary coolant escapes into the containment structure, certain noble gas nuclides become airborne, and subsequently decay into particulate nuclides. One of the most common of these pairs is ^{88}Kr and ^{88}Rb; the latter is detected by the CPAM. Relating the observed CPAM response to the ^{88}Rb back to a leakage rate from the primary system is far from trivial.

The regulatory basis for this CPAM application is found in 10CFR50:

For use in the USA, standard 10 CFR 50, Appendix A, "General Design Criteria for Nuclear Power Plants," Criterion 30, "Quality of reactor coolant pressure boundary," requires that means be provided for detecting and, to the extent practical, identifying the location of the source of reactor coolant leakage. The specific attributes of the reactor coolant leakage detection systems are outlined in Regulatory Positions 1 through 9 of Regulatory Guide 1.45.

For use in the USA, standard 10 CFR 50.36, "Technical Specifications," paragraph (c)(2)(ii)(A), specifies that a Limiting Condition for Operation be established for installed instrumentation that is used to detect and indicate in the control room a significant abnormal degradation of the reactor coolant pressure boundary. This instrumentation is required by Specification 3.4.15, "RCS Leakage Detection Instrumentation."

Step changes in reactor coolant leakage can be detected with moving filter media to sat-isfy the quantitative requirements of USNRC Regulatory Guide 1.45. The mathematical method is highly detailed and it focuses on time-dependent viewable collected activity, rather than concentration, as f(t). The method, among other features, yields the desired fixed-filter degenerate case (filter paper velocity = 0.) The method was first put into use in the 1990s at a nuclear power plant in the United States. Though originally derived for dominant Kr-88/Rb-88 in leaked reactor coolant, it has been expanded to include Xe-138/Cs-138 and can be modified by replication to include any N similar pairings. Further refinements to the mathematical methodology have been made by the inventor; these developments obviate the described patented collimator apparatus for making quantitative assessment of leak rate step change when rectangular collection grids are employed.

Some CPAM Application Considerations

Importance of Nuclide Half-life

The response of the monitor is sensitive to the half-life of the nuclide being collected and measured. It is useful to define a "long-lived" (LL) nuclide to have negligible decay during the measurement interval. On the other hand, if the decay cannot be ignored, the nuclide is considered "short-lived" (SL). In general, for the monitor response models discussed below, the LL response can be obtained from the SL response by taking limits of the SL equation as the decay constant approaches zero. If there is any question about which response model to use, the SL expressions will *always* apply; however, the LL equations are considerably simpler and so should be used when there is no question about the half-life (e.g., ^{137}Cs is LL).

Ratemeter

The output of the radiation detector is a random sequence of pulses, usually processed by some form of "ratemeter," which continuously estimates the rate at which the detector is responding to the radioactivity deposited on the filter medium. There are two fundamental types of ratemeters, analog and digital. The ratemeter output is called the countrate, and it varies with time.

Ratemeters of both types have the additional function of "smoothing" the output countrate estimate, i.e., reducing its variability. (This process is more correctly termed "filtering.") Ratemeters must make a tradeoff between this necessary variance reduction and their response time; a smooth output (small variance) will tend to lag behind an increase in the true pulse rate. The significance of this lag depends on the application of the monitor.

Ambient Background

Even when the filter medium is clean, that is, before the pump is started that pulls the air through the filter, the detector will respond to the ambient "background" radiation in the vicinity of the monitor. The countrate that results from deposited radioactivity is called the "net" countrate, and is obtained by subtracting this background countrate from the dynamically-varying countrate that is observed once the pump is started. The background is usually assumed to be constant.

Integration Time

The countrate of the monitor varies dynamically, so that a measurement time interval must be specified. Also, these are integrating devices, meaning that some finite time is required to accumulate radioactivity onto the filter medium. The input to the monitor is, in general, a time-dependent concentration in air of the specified nuclide. However, for the calculations given below, this concentration will be held constant over that interval.

Constant-concentration Time Limitation

Since concentrations resulting from physical events tend to vary with time, due to dilution processes and/or a nonconstant source term (airborne radioactivity emission rate), it is not realistic to hold the concentration constant for significant lengths of time. Thus, measurement intervals on the order of several hours are not plausible for the purposes of these calculations.

Parent-progeny; RnTn

There are situations in which a nuclide deposited on the CPAM filter decays into another nuclide, and that second nuclide remains on the filter. This "parent-progeny" or decay chain situation is especially relevant to so-called "radon-thoron" (RnTn) or natural airborne radioactivity. The mathematical treatment described in this article does not consider this situation, but it can be treated using matrix methods.

Multiple Nuclides; Superposition

Another issue is the fact that in a power reactor context it would be unusual for a CPAM to be collecting only a single particulate nuclide; more likely there would be a mixture of fission product and activation product nuclides. The modeling discussed in this article considers only one nuclide at a time. However, since the radiation emitted by each nuclide is independent of the others, so that the nuclides present on the filter medium do not interact with each other, the monitor response is the linear combination of the individual responses. Thus the overall CPAM response to a mixture is just the superposition (i.e., the sum) of the individual responses.

Detector Type

CPAMs use either a Geiger tube, for "gross beta-gamma" counting, or a NaI(Tl) crystal, often for simple single-channel gamma spectroscopy. (In this context, "gross" means a measurement that does not attempt to find the specific nuclides in the sample.) Plastic scintillators are also popular. Essentially, in power reactor applications, beta and gamma are the radiations of interest for particulate monitoring.

In other fuel-cycle applications, such as nuclear reprocessing, alpha detection is of interest. In those cases, the interference from other isotopes such as RnTn is a major problem, and more sophisticated analysis, such as the use of HPGe detectors and multichannel analyzers, are used where spectral information, such as is used for Radon compensation, is required.

Radioiodine (especially ^{131}I) monitoring is often done using a particulate-monitor setup, but with an activated charcoal collection medium, which can adsorb some iodine vapors as well as particulate forms. Single-channel spectroscopy is usually specified for iodine monitors.

Dynamic Response of CPAMs

Detailed mathematical models that describe the dynamic, time-dependent countrate response of these monitors in a very general manner are presented in and will not be repeated here. For the purpose of this article, a few useful results from that paper will be summarized. The objective is to predict the net countrate of a CPAM for a single, specific manmade nuclide, for a given set of conditions. That predicted response can be compared to the expected background and/or interferences (nuclides other than the one sought), to assess the monitor's detection capability. The response predictions can also be used to calculate alarm setpoints that correspond to appropriate limits (such as those in 10CFR20) on the concentration of airborne radioactivity in the sampled air.

Model Parameters

The parameters used in these models are summarized in this list:

- Time interval (t); time; measured from start of concentration step

- Concentration (Q_o); activity / volume; assumed constant over the interval

- Decay constant (λ); 1 / time; for the specified nuclide

- Media collection/retention efficiency (φ); implicitly includes line loss

- Window length or radius (L or R); length; consistent units with v

- Filter speed (v); length / time; length has same units as L or R

- Flow rate (F_m); volume / time; assumed constant over the interval

- Detection efficiency (ε); counts / disintegration; implicitly includes emission abundance

"Line loss" refers to the losses of particulate matter in transit from a sampling point to the monitor; thus the concentration measured would be somewhat lower than that in the original sampled air. This factor is meant to compensate for these losses. Sampling lines are specifically designed to minimize these losses, for example, by making bends gradual as opposed to right-angled. These lines (pipes) are needed since in many applications the CPAM cannot be physically located directly in the sampled air volume, such as a nuclear power plant's main stack, or the ventilation air intake for the plant control room.

"Emission abundance" refers to the fact that the disintegration of any given nucleus of the isotope of interest in the CPAM analysis may not result in the emission of the radiation being detected (e.g., a beta particle or gamma ray). Thus, overall there will be some fraction of the disintegrations that emit the radiation of interest (e.g. the 662 keV gamma ray of ^{137}Cs is emitted in about 85% of the disintegrations of ^{137}Cs nuclei).

Fixed-filter Model

The response models are based on the consideration of the sources and losses of the deposited radioactivity on the filter medium. Taking the simplest case, the FF monitor, this leads to a differential equation which expresses the rate of change of the monitor countrate:

$$\frac{d\dot{C}_{FF}}{dt} = \varepsilon k F_m \phi Q(t) - \lambda \dot{C}_{FF}$$

The first term accounts for the source of radioactivity from the sampled air, and the second term is the loss due to the decay of that radioactivity. A convenient way to express the solution to this equation uses the scalar convolution integral, which results in

$$\dot{C}_{FF}(t) = \varepsilon k F_m \phi \exp(-\lambda t) \int_0^t Q(\tau) \exp(\lambda \tau) d\tau + \dot{C}_0 \exp(-\lambda t)$$

The last term accounts for any initial activity on the filter medium, and is usually set to zero (clean filter at time zero). The initial countrate of the monitor, before the concentration transient begins, is only that due to ambient background. If radon progeny are present, they are assumed to be at equilibrium and generating a constant countrate that adds to the ambient background's countrate.

Various solutions for the time-dependent FF countrate follow directly, once a concentration time-dependence $Q(t)$ has been specified. Note that the monitor flowrate F_m is assumed constant; if it isn't, and its time-dependence is known, then that $F_m(t)$ would need to be placed inside the integral. Also note that the time variable in all the models is measured from the instant the concentration in the sampled air begins to increase.

Moving-filter Models

For the moving-filter CPAMs, the above expression is a starting point, but the models are considerably more complicated, due to (1) the loss of material as the filter medium moves away from the detector's field of view and (2) the differing lengths of time that parts of the filter medium have been exposed to the sampled air. The basic modeling approach is to break down the deposition regions into small differential areas and then consider how long each such area receives radioactive material from the air.

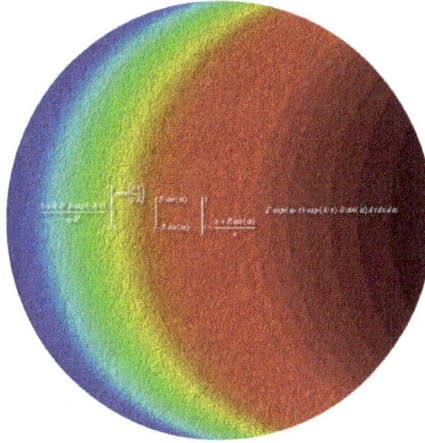

Circular-window moving filter monitor; deposited radioactivity isoactivity contours, after transit time, constant input concentration.

The resulting expressions are integrated across the deposition region to find the overall response. The RW solution consists of two double integrals, while the CW response solution consists of three triple integrals. A very important consideration in these models is the "transit time," which is the time required for a differential area to traverse the window along its longest dimension. As a practical matter, the transit time is the time required for *all* differential elements that were in the deposition window at time zero to leave the window.

This figure shows contours of constant activity on a CW deposition area, after the transit time has expired. The filter moves from left to right, and the activity increases from left to right. The differential areas on the diameter have been in the deposition window the longest, and at the far right, have been in the window, accumulating activity, for the full transit time.

Finally, to illustrate the complexity of these models, the RW response for time less than the transit time is

$$\dot{C}_{RW}(t) = \frac{\varepsilon k F_m \phi \exp(-\lambda t)}{L} \left[\int_0^{vt} \int_{t-\left(\frac{x}{v}\right)}^{t} Q(\tau)\exp(\lambda\tau)d\tau dx + \int_{vt}^{L}\int_0^{t} Q(\tau)\exp(\lambda\tau)d\tau dx \right]$$

and, also, one of the CW triple integrals is superimposed on the contour plot.

Selected CPAM Response Models: Constant Concentration

In these equations, k is a conversion constant for units reconciliation. Again, a very important parameter for moving-filter monitors is the "transit time" (T), which is the window length (or diameter) divided by the filter tape speed v. The countrate is denoted by $\dot{C}$.

Fixed-filter (FF), any half-life

$$\dot{C}_{FF}(t) = \varepsilon k F_m \phi Q_0 \frac{1-\exp(-\lambda t)}{\lambda}$$

Fixed-filter (FF), long-lived (LL)

$$\dot{C}_{FF}(t) = \varepsilon k F_m \phi Q_0 t$$

Rectangular window (RW), time less than transit time T, any half-life

$$\dot{C}_{RW}(t) = \frac{\varepsilon k F_m \phi Q_0}{\lambda^2}\frac{v}{L}[\lambda t - 1 + \exp(-\lambda t)] + \frac{\varepsilon k F_m \phi Q_0}{\lambda}\left(1 - \frac{vt}{L}\right)[1-\exp(-\lambda t)]$$

Rectangular window (RW), time less than transit time T, LL

$$\dot{C}_{RW}(t) = \varepsilon k F_m \phi Q_0 \left(t - \frac{vt^2}{2L}\right)$$

Note that as v approaches zero, these RW equations reduce to the FF solutions.

Rectangular window (RW), time greater than or equal to transit time T, any half-life

$$\dot{C}_{RW}(t) = \varepsilon k F_m \phi Q_0 \left\{\frac{1}{\lambda} - \frac{v}{\lambda^2 L}\left[1 - \exp\left(-\lambda\frac{L}{v}\right)\right]\right\}$$

Rectangular window (RW), time greater than or equal to transit time T, LL

$$\dot{C}_{RW}(t) = \varepsilon k F_m \phi Q_0 \frac{L}{2v}$$

Circular window (CW) responses

These response-model equations are quite complicated and some involve a nonelementary integral; the exact solutions can be found here. It is shown here, however, that a reasonable approximation for predicting the CW response can be obtained by using the RW equations above, with an "adjusted" window length L_{CW} used in each occurrence of the parameter L, except that the CW transit time T_{CW} is found from 2R / v, *not* from using L_{CW} as given here in the T_{RW} relation L / v. Thus,

$$L_{CW} = \frac{16R}{3\pi} \qquad T_{CW} = \frac{2R}{v} \neq \frac{L_{CW}}{v} \qquad T_{RW} = \frac{L}{v}$$

Example CPAM Response Plots

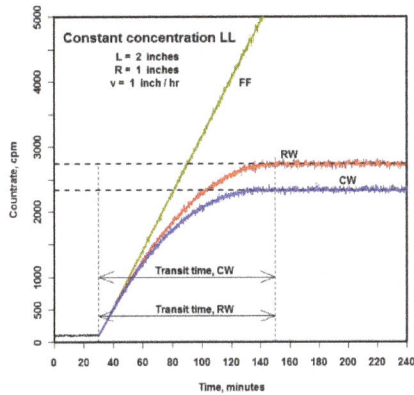

CPAM responses, constant concentration of LL activity. Transit time 120 min.

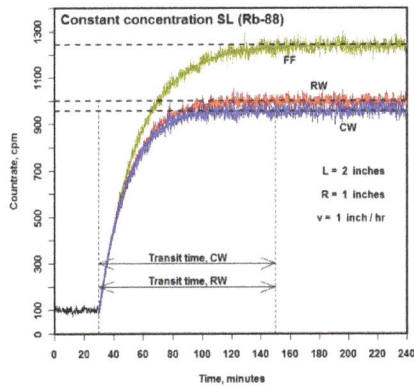

CPAM responses, constant concentration of SL activity (Rb-88). Transit time 120 min.

These plots show the predicted CPAM countrate responses for these parameter settings: Detection efficiency, 0.2; Flowrate, 5 cubic feet per minute (cfm); Collection efficiency, 0.7; Constant concentration, 1E-09 Ci/cc; Rectangular window length, 2 inches; Circular window radius, 1 inch; Media (tape) speed, 1 inch/hour. The concentration instantly steps up to its constant value when the time reaches 30 minutes, and there is a 100 count per minute (cpm) constant background. Note: A microcurie (Ci) is a measure of the disintegration rate, or activity, of a radioactive source; it is 2.22E06 disintegrations per minute.

In the LL plot, note that the FF countrate continues to increase. This is because there is no significant loss of radioactivity from the filter medium. The RW and CW monitors, on the other hand, approach a limiting countrate and the monitor response remains constant as long as the input concentration remains constant.

For the SL plot, all three monitor responses approach a constant level. For the FF monitor, this is due to the source and loss terms becoming equal; since ^{88}Rb has a half-life of about 18 minutes, the loss of radioactive material from the filter medium is significant. This loss also happens on the RW and CW monitors, but there, the loss due to the filter movement also plays a role.

In both plots, Poisson "noise" is added and a constant-gain digital filter is applied, emulating the countrate responses as they would be observed on a modern CPAM. The horizontal dotted lines are the limiting countrates calculated from the equations given in the previous section.

Also in both plots the transit times are indicated; note that these times are measured from the *start of the concentration,* at time 30 minutes, *not* from the arbitrary time zero of the plots. In these example graphs, the length of the RW and the diameter of the CW are equal; if they were not equal, then the transit times would not be equal.

The Inverse Problem: Estimating a Concentration From the Observed Response

Having mathematical models that can predict the CPAM response, i.e., the monitor's output, for a defined input (airborne radioactive material concentration), it is natural to ask whether the process can be "inverted." That is, given an observed CPAM *output,* is it possible to estimate the *input* to the monitor?

A Misleading "Quantitative Method" for Moving-filter CPAMs

A number of approaches to this inverse problem are addressed in detail in. Each method has its advantages and disadvantages, as one might expect, and a method that might work well for a fixed-filter monitor may be useless for a moving-filter monitor (or vice versa).

One important conclusion from this paper is that for all practical purposes *moving-filter monitors are not usable for quantitative estimation of a time-dependent concentration.* The only moving-filter method that has been used historically involves a constant-concentration, LL assumption, which leads to the RW expression:

$$\dot{C}_{RW} = \varepsilon k F_m \phi Q_0 \frac{T}{2} \quad \Rightarrow \quad \hat{Q}_0 = \frac{2v\dot{C}_{RW}}{\varepsilon k F_m \phi L}$$

or for CW,

$$\dot{C}_{CW} = \varepsilon k F_m \phi Q_0 \frac{8R}{3\pi v} \quad \Rightarrow \quad \hat{Q}_0 = \frac{3\pi v \dot{C}_{CW}}{8R\varepsilon k F_m \phi}$$

Thus, a concentration estimate is available *only after the transit time T has expired;* in most CPAM applications this time is on the order of several (e.g., 4) hours. Whether it is reasonable to assume that the concentration will stay constant for this length of time, and to further assume that only long-lived nuclides are present, is at least debatable, and it is arguable that in many practical situations these assumptions are not realistic.

For example, in power reactor leak detection applications, as mentioned in the first section of this article, CPAMs are used, and a primary nuclide of interest is [88]Rb, which is far from long-lived (half-life 18 minutes). Also, in the dynamic environment of a reactor containment building the [88]Rb concentration would not be expected to remain constant on a time scale of hours, as required by this measurement method.

However, realistic or not, it has for decades been the practice of CPAM vendors to provide a set of curves (graphs) based on the expressions above. Such graphs have concentration on the vertical axis, and net countrate on the horizontal axis. There often is a family of curves, parameterized on the detection efficiency (or labeled as to specific nuclides). The implication in providing these graphs is that one is to observe a net countrate, at any time, enter the graph at this value, and read off the concentration that exists at that time. To the contrary, unless the time is greater than the transit time T, the nuclide of interest is long-lived, and the concentration is constant over the entire interval, this process will lead to incorrect concentration estimates.

Quantitative Methods for CPAM Applications

As discussed in the referenced paper, there are at least 11 possible quantitative methods for estimating the concentration or quantities derived from it. The "concentration" may only be at a specific time, or it might be an average over some time interval; this averaging is perfectly acceptable in some applications. In a few cases, the time-dependent concentration itself can be estimated. These various methods involve the countrate, the time derivative of the countrate, the time integral of the countrate, and various combinations of these.

The countrate is, as mentioned above, developed from the raw detector pulses by either an analog or digital ratemeter. The integrated counts are easily obtained simply by accumulating the pulses in a "scaler" or, in more modern implementations, in software. Estimating the rate of change (time derivative) of the countrate is difficult to do with any reasonable precision, but modern digital signal processing methods can be used to good effect.

It turns out that it is very useful to find the time integral of the concentration, as opposed to estimating the time-dependent concentration itself. It is essential to consider this choice for any CPAM application; in many cases the integrated concentration is not only more useful in a radiological protection sense, but is also more readily accomplished, since estimating a concentration in (more or less) real-time is difficult.

For example, the total activity released from a plant stack over a time interval is

$$R_{stack}(\eta) = \int_0^{\eta} Q(\tau) F_{stack}(\tau) d\tau$$

Then, for a fixed-filter monitor, assuming a constant stack and monitor flowrate, it can be shown that

$$R_{stack}(\eta) = \frac{F_{stack}\left[\dot{C}(\eta) + \lambda\int_{0}^{\eta}\dot{C}(\tau)d\tau\right]}{\varepsilon k F_{m}\phi}$$

so that the release is a function of both the countrate and integrated counts. This approach was implemented at the SM-1 Nuclear Power Plant in the late 1960s, for estimating the releases of episodic containment purges, with a predominant, and strongly time-varying, nuclide of ^{88}Rb. For a LL nuclide, the integral term vanishes, and the release depends only on the attained countrate. A similar equation applies for the occupational exposure situation, replacing the stack flowrate with a worker's breathing rate.

An interesting subtlety to these calculations is that the time in the CPAM response equations is measured from the *start* of a concentration transient, so that some method of detecting the resulting change in a noisy countrate must be developed. Again, this is a good application for statistical signal processing that is made possible by the use of computing power in modern CPAMs.

Which of these 11 methods to use for the applications discussed previously is not especially obvious, although there are some candidate methods that logically would be used in some applications and not in others. For example, the response time of a given CPAM quantitative method may be far too slow for some applications, and perfectly reasonable for others. The methods have varying sensitivities (detection capabilities; how small a concentration or quantity of radioactivity can *reliably* be detected) as well, and this must enter into the decision.

CPAM Calibration

The calibration of a CPAM usually includes: (1) choosing a quantitative method; (2) estimating the parameters needed to implement that method, notably the detection efficiency for specified nuclides, as well as the sampling line loss and collection efficiency factors; (3) estimating, under specified conditions, the background response of the instrument, which is needed for calculating the detection sensitivity. This sensitivity is often called the *minimum detectable concentration* or MDC, assuming that a concentration is the quantity estimated by the selected quantitative method.

What is of interest for the MDC is the variability (not the level) of the CPAM background countrate. This variability is measured using the standard deviation; care must be taken to account for bias in this estimate due to the autocorrelation of the sequential monitor readings. The autocorrelation bias can make the calculated MDC significantly *smaller* than is actually the case, which in turn makes the monitor appear to be capable of reliably detecting smaller concentrations than it in fact can.

An uncertainty analysis for the estimated quantity (concentration, release, uptake) is also part of the calibration process. Other performance characteristics can be part of this process, such as estimating response time, estimating the effect of temperature changes on the monitor response, and so on.

Table of Radiation Measurement Quantities

This is given to show context of US and SI units.

Quantity	Name	Symbol	Unit	Year	System
Exposure (X)	röntgen	R	esu / 0.001293 g of air	1928	non-SI
Absorbed dose (D)			erg•g⁻¹	1950	non-SI
	rad	rad	100 erg•g⁻¹	1953	non-SI
	gray	Gy	J•kg⁻¹	1974	SI
Activity (A)	curie	Ci	3.7×10^{10} s⁻¹	1953	non-SI
	becquerel	Bq	s⁻¹	1974	SI
Dose equivalent (H)	röntgen equivalent man	rem	100 erg•g⁻¹	1971	non-SI
	sievert	Sv	J•kg⁻¹	1977	SI
Fluence (Φ)	(reciprocal area)		cm⁻² or m⁻²	1962	SI (m⁻²)

Although the United States Nuclear Regulatory Commission permits the use of the units curie, rad, and rem alongside SI units, the European Union European units of measurement directives required that their use for "public health ... purposes" be phased out by 31 December 1985.

Ionization Chamber

The ionization chamber is the simplest of all gas-filled radiation detectors, and is widely used for the detection and measurement of certain types of ionizing radiation; X-rays, gamma rays and beta particles. Conventionally, the term "ionization chamber" is used exclusively to describe those detectors which collect all the charges created by *direct ionization* within the gas through the application of an electric field. It only uses the discrete charges created by each interaction between the incident radiation and the gas, and does not involve the gas multiplication mechanisms used by other radiation instruments, such as the Geiger-Müller counter or the proportional counter.

Ion chambers have a good uniform response to radiation over a wide range of energies and are the preferred means of measuring high levels of gamma radiation. They are widely used in the nuclear power industry, research labs, radiography, radiobiology, and environmental monitoring.

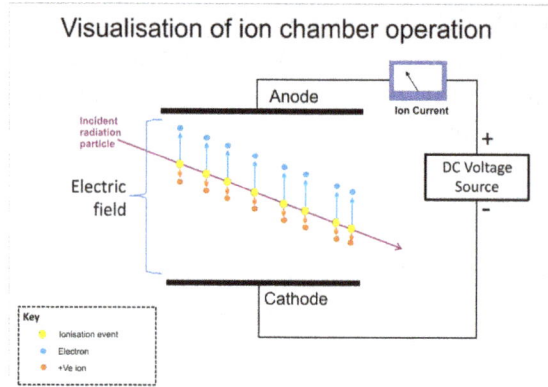

Schematic diagram of parallel plate ion chamber, showing drift of ions. Electrons typically drift 1000 times faster than positive ions due to their much smaller mass.

Principle of Operation

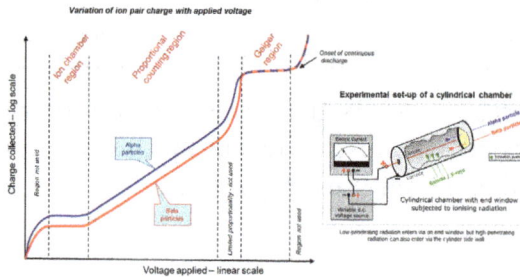

Plot of ion current against voltage for a wire cylinder gaseous radiation detector. The ion chamber uses the lowest usable detection region.

An ionization chamber measures the charge from the number of ion pairs created within a gas caused by incident radiation. It consists of a gas-filled chamber with two electrodes; known as anode and cathode. The electrodes may be in the form of parallel plates (Parallel Plate Ionization Chambers: PPIC), or a cylinder arrangement with a coaxially located internal anode wire.

A voltage potential is applied between the electrodes to create an electric field in the fill gas. When gas between the electrodes is ionized by incident ionizing radiation, ion-pairs are created and the resultant positive ions and dissociated electrons move to the electrodes of the opposite polarity under the influence of the electric field. This generates an ionization current which is measured by an electrometer circuit. The electrometer must be capable of measuring the very small output current which is in the region of femtoamperes to picoamperes, depending on the chamber design, radiation dose and applied voltage.

Each ion pair created deposits or removes a small electric charge to or from an electrode, such that the accumulated charge is proportional to the number of ion pairs created, and hence the radiation dose. This continual generation of charge produces an ionization current, which is a measure of the *total* ionizing dose entering the chamber. However, the chamber cannot discriminate between radiation types (beta or gamma) and cannot produce an energy spectrum of radiation.

The electric field also enables the device to work continuously by mopping up electrons, which prevents the fill gas from becoming saturated, where no more ions could be collected, and by preventing the recombination of ion pairs, which would diminish the ion current. This mode of operation is referred to as "current" mode, meaning that the output signal is a continuous current, and not a pulse output as in the cases of the Geiger-Müller tube or the proportional counter.

Referring to the accompanying ion pair collection graph, it can be seen that in the "ion chamber" operating region the collection of ion pairs is effectively constant over a range of applied voltage, as due to its relatively low electric field strength the ion chamber does not have any "multiplication effect". This is in distinction to the Geiger-Müller tube or the proportional counter whereby secondary electrons, and ultimately multiple avalanches, greatly amplify the original ion-current charge.

Chamber Types and Construction

The following chamber types are commonly used.

Free-air Chamber

This is a chamber freely open to atmosphere, where the fill gas is ambient air. The domestic smoke detector is a good example of this, where a natural flow of air through the chamber is necessary so that smoke particles can be detected by the change in ion current. Other examples are applications where the ions are created outside the chamber but are carried in by a forced flow of air or gas.

Chamber Pressure

Vented Chamber

These chambers are normally cylindrical and operate at atmospheric pressure, but to prevent ingress of moisture a filter containing a desiccant is installed in the vent line. This is to stop moisture building up in the interior of the chamber, which would otherwise be introduced by the "pump" effect of changing atmospheric air pressure. These chambers have a cylindrical body made of aluminium or plastic a few millimetres thick. The material is selected to have an atomic number similar to that of air so that the wall is said to be "air equivalent" over a range of radiation beam energies. This has the effect of ensuring the gas in the chamber is acting as though it were a portion of an infinitely large gas volume, and increases the accuracy by reducing interactions of gamma with the wall material. The higher the atomic number of the wall material, the greater the chance of interaction. The wall thickness is a trade-off between maintaining the air effect with a thicker wall, and increasing sensitivity by using a thinner wall. These chambers often have an end window made of material thin enough, such as mylar, so that beta particles can enter the gas volume. Gamma radiation enters both through the end window and the side walls. For hand-held instruments the wall thickness is made as uniform as possible to reduce photon directionality though any beta window response is obviously highly directional. Vented chambers are susceptible to small changes in efficiency with air pressure and correction factors can be applied for very accurate measurement applications.

Sealed Low Pressure Chamber

These are similar in construction to the vented chamber but are sealed and operate at or around atmospheric pressure. They contain a special fill gas to improve detection efficiency as free electrons are easily captured in air-filled vented chambers by neutral oxygen which is electronegative, to form negative ions. These chambers also have the advantage of not requiring a vent and desiccant. The beta end window limits the differential pressure from atmospheric pressure that can be tolerated, and common materials are stainless steel or titanium with a typical thickness of 25 μm.

High Pressure Chamber

The efficiency of the chamber can be further increased by the use of a high pressure gas. Typically a pressure of 8-10 atmospheres can be used, and various noble gases are employed. The higher pressure results in a greater gas density and thereby a greater chance of collision with the fill gas and ion pair creation by incident radiation. Because of the increased wall thickness required to withstand this high pressure, only gamma radiation can be detected.These detectors are used in survey meters and for environmental monitoring.

Chamber Shape

Thimble Chamber

Most commonly used for radiation therapy measurements is a cylindrical or "thimble" chamber. The active volume is housed within a thimble shaped cavity with an inner conductive surface (cathode) and a central anode. A bias voltage applied across the cavity collects ions and produces a current which can be measured with an electrometer.

Parallel-plate Chambers

Parallel-plate chambers are shaped like a small disc, with circular collecting electrodes separated by a small gap, typically 2mm or less. The upper disc is extremely thin, allowing for much more accurate near-surface dose measurements than are possible with a cylindrical chamber.

Monitor Chambers

Monitor chambers are typically parallel plate ion chambers which are placed in radiation beams to continuously measure the beam's intensity. For example, within the head of linear accelerators used for radiotherapy, multi-cavity ionization chambers can measure the intensity of the radiation beam in several different regions, providing beam symmetry and flatness information.

Research and Calibration Chambers

Ionization chamber made by Pierre Curie, c 1895-1900

Early versions of the Ion chamber were used by Marie and Pierre Curie in their original work in isolating radioactive materials. Since then the ion chamber has been a widely used tool in the laboratory for research and calibration purposes. To do this a wide variety of bespoke chamber shapes, some using liquids as the ionized medium, have been evolved and used. Ion chambers are used by national laboratories to calibrate primary standards, and also to transfer these standards to other calibration facilities.

Historical Chambers

Condenser Chamber

The condenser chamber has a secondary cavity within the stem which acts as a capacitor. When this capacitor is fully charged, any ionization within the thimble counteracts this charge, and the change in charge can be measured. They are only practical for beams with energy of 2 MeV or less, and high stem leakage makes them unsuited to precise dosimetry.

Extrapolation Chamber

Similar in design to a parallel plate chamber, the upper plate of an extrapolation chamber can be lower using micrometer screws. Measurements can be taken with different plate spacing and extrapolated to a plate spacing of zero, i.e. the dose without the chamber.

Instrument Types

Hand Held

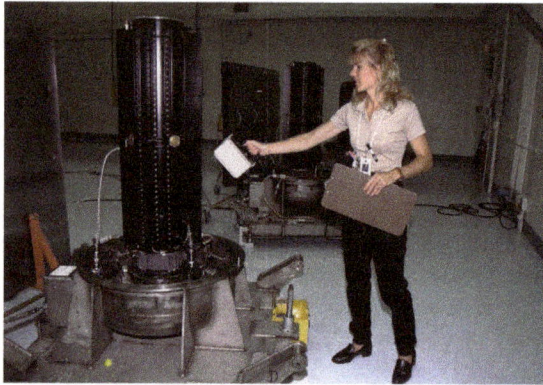

Hand-held integral ion chamber survey meter in use

View of sliding beta shield on integral hand held instrument

Ion chambers are widely used in hand held radiation survey meters to measure beta and gamma radiation. They are particularly preferred for high dose rate measurements and for gamma radiation they give good accuracy for energies above 50-100 keV.

There are two basic configurations; the "integral" unit with the chamber and electronics in the same case, and the "two-piece" instrument which has a separate ion chamber probe attached to the electronics module by a flexible cable.

The chamber of the integral instrument is generally at the front of the case facing downwards, and for beta/gamma instruments there is a window in the bottom of the casing. This usually has a sliding shield which enables discrimination between gamma and beta radiation. The operator closes the shield to exclude beta, and can thereby calculate the rate of each radiation type.

Some hand held instruments generate audible clicks similar to that produced by a G-M counter to assist operators, who use the audio feedback in radiation survey and contamination checks. As the ion chamber works in current mode, not pulse mode, this is synthesised from the radiation rate.

Installed

For industrial process measurements and interlocks with sustained high radiation levels, the ion chamber is the preferred detector. In these applications only the chamber is situated in the measurement area, and the electronics are remotely situated to protect them from radiation and connected by a cable. Installed instruments can be used for measuring ambient gamma for personnel protection and normally sound an alarm above a preset rate, though the Geiger-Müller tube instrument is generally preferred where high levels of accuracy are not required.

General Precautions in Use

Moisture is the main problem that affects the accuracy of ion chambers. The chamber's internal volume must be kept completely dry, and the vented type uses a desiccant to help with this. Because of the very low currents generated, any stray leakage current must be kept to a minimum in order to preserve accuracy. Invisible hygroscopic moisture on the surface of cable dielectrics and connectors can be sufficient to cause a leakage current which will swamp any radiation-induced ion current. This requires scrupulous cleaning of the chamber, its terminations and cables, and subsequent drying in an oven. "Guard rings" are generally used as a design feature on higher voltage tubes to reduce leakage through or along the surface of tube connection insulators which can require a resistance in the order of 10^{13} Ω.

For industrial applications with remote electronics, the ion chamber is housed in a separate enclosure which provides mechanical protection and contains a desiccant to remove moisture which could affect the termination resistance.

In installations where the chamber is a long distance from the measuring electronics, readings can be affected by external electromagnetic radiation acting on the cable. To overcome this a local converter module is often used to translate the very low ion chamber currents to a pulse train or data signal related to the incident radiation. These are immune to electromagnetic effects.

Applications

Nuclear Industry

Ionization chambers are widely used in the nuclear industry as they provide an output that is proportional to radiation dose They find wide use in situations where a constant high dose rate is being measured as they have a greater operating lifetime than standard Geiger-Müller tubes, which suffer from gas break down and are generally limited to a life of about 10^{11} count events. Additionally, the Geiger-Müller tube cannot operate above about 10^4 counts per second, due to dead time effects, whereas there is no similar limitation on the ion chamber.

Smoke Detectors

The ionization chamber has found wide and beneficial use in smoke detectors. In a smoke detector, ambient air is allowed to freely enter the ionization chamber. The chamber contains a small amount of americium-241, which is an emitter of alpha particles which produce a constant ion current. If smoke enters the detector, it disrupts this current because ions strike smoke particles and are neutralized. This drop in current triggers the alarm. The detector also has a reference chamber which is sealed but is ionized in the same way. Comparison of the ion currents in the two chambers allows compensation for changes due to air pressure, temperature, or the ageing of the source.

Medical Radiation Measurement

In medical physics and radiotherapy, ionization chambers are used to ensure that the dose delivered from a therapy unit or radiopharmaceutical is what is intended. The devices used for radiotherapy are called "reference dosimeters", while those used for radiopharmaceuticals are called [[radioisotope dose calibrators] - an inexact name for radionuclide radioactivity calibrators, which are used for measurement of radioactivity but not absorbed dose]. A chamber will have a calibration factor established by a national standards laboratory such as ARPANSA in Australia or the NPL in the UK, or will have a factor determined by comparison against a transfer standard chamber traceable to national standards at the user's site,.

Guidance on Application Use

In the United Kingdom the HSE has issued a user guide on selecting the correct radiation measurement instrument for the particular application concerned. This covers all radiation instrument technologies, and is a useful comparative guide to the use of ion chamber instruments.

Gaseous Ionization Detectors

Gaseous ionization detectors are radiation detection instruments used in particle physics to detect the presence of ionising particles, and in radiation protection applications to measure ionizing radiation.

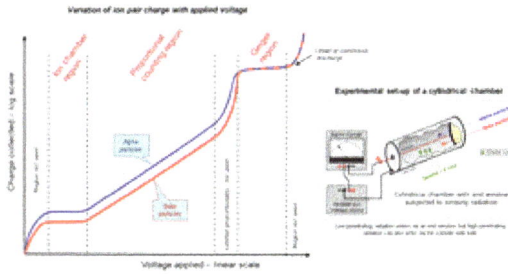

Plot of variation of ion pair current against applied voltage for a wire cylinder gaseous radiation detector.

They use the ionising effect of radiation upon a gas-filled sensor. If a particle has enough energy to ionize a gas atom or molecule, the resulting electrons and ions cause a current flow which can be measured.

Gaseous ionisation detectors form an important group of instruments used for radiation detection and measurement. This article gives a quick overview of the principal types, and more detailed information can be found in the articles on each instrument. The accompanying plot shows the variation of ion pair generation with varying applied voltage for constant incident radiation. There are three main practical operating regions, one of which each type utilises.

Types

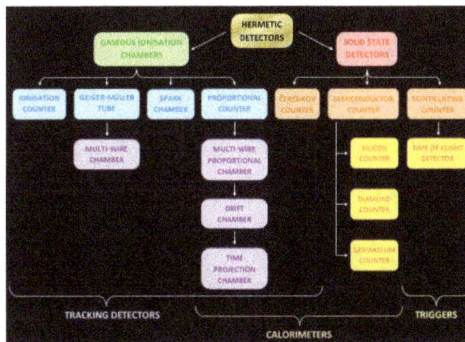

Families of ionising radiation detectors

The three basic types of gaseous ionization detectors are:

- ionization chambers

- proportional counters

- Geiger-Müller tubes

All of these have the same basic design of two electrodes separated by air or a special fill gas, but each uses a different method to measure the total number of ion-pairs that are collected. The strength of the electric field between the electrodes and the type and pressure of the fill gas determines the detector's response to ionizing radiation.

Ionization Chamber

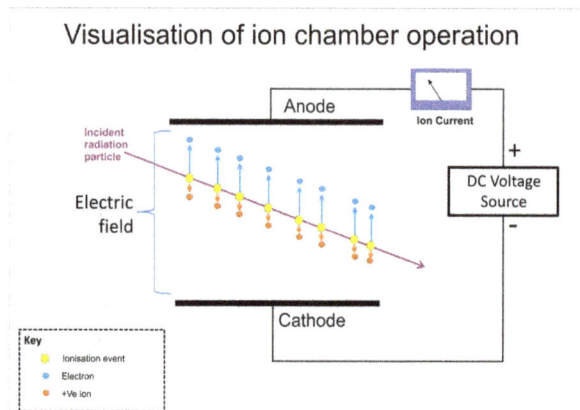

Schematic diagram of ion chamber, showing drift of ions. Electrons typically drift 1000 times faster than positive ions due to their much smaller mass.

Ionization chambers operate at a low electric field strength, selected such that no gas multiplication takes place. The ion current is generated by the creation of "ion pairs", consisting of an ion and an electron. The ions drift to the cathode whilst free electrons drift to the anode under the influence of the electric field. This current is independent of the applied voltage if the device is being operated in the "ion chamber region". Ion chambers are preferred for high radiation dose rates because they have no "dead time"; a phenomenon which affects the accuracy of the Geiger Muller tube at high dose rates.

The advantages are:

- Good uniform response to gamma radiation and give an accurate overall dose reading

- Will measure very high radiation rates

- Sustained high radiation levels do not degrade fill gas

The disadvantages are:

- Very low electronic output requiring sophisticated electrometer circuit

- Operation and accuracy easily affected by moisture

Proportional Counter

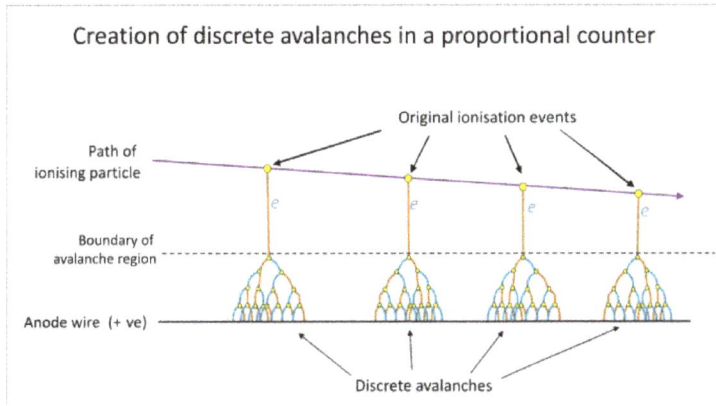

The generation of discrete Townsend avalanches in a proportional counter.

Proportional counters operate at a slightly higher voltage, selected such that discrete avalanches are generated. Each ion pair produces a single avalanche so that an output current pulse is generated which is proportional to the energy deposited by the radiation. This is in the "proportional counting" region. The term "gas proportional detector" (GPD) is generally used in radiometric practice, and the property of being able to detect particle energy is particularly useful when using large area flat arrays for alpha and beta particle detection and discrimination, such as in installed personnel monitoring equipment. The Wire chamber is a multi-electrode form of proportional counter used as a research tool.

The advantages are:

- Can measure energy of radiation and provide spectrographic information

- Can discriminate between alpha and beta particles

- Large area detectors can be constructed

The disadvantages are:

- Anode wires delicate and can lose efficiency in gas flow detectors due to deposition

- Efficiency and operation affected by ingress of oxygen into fill gas

- Measurement windows easily damaged in large area detectors

Geiger-müller Tube

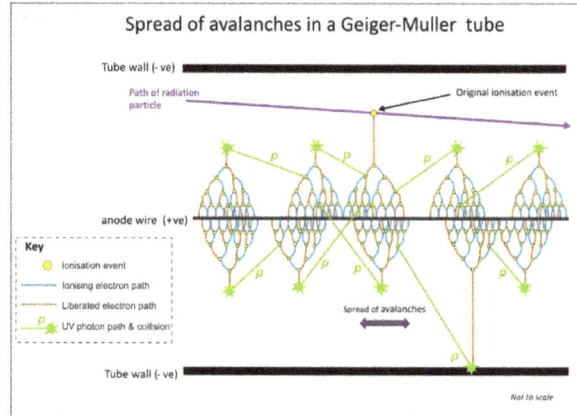

Visualisation of the spread of Townsend avalanches by means of UV photons

Geiger-Müller tubes are the primary components of Geiger counters. They operate at an even higher voltage, selected such that each ion pair creates an avalanche, but by the emission of UV photons, multiple avalanches are created which spread along the anode wire, and the adjacent gas volume ionizes from as little as a single ion pair event. This is the "Geiger region" of operation. The current pulses produced by the ionising events are passed to processing electronics which can derive a visual display of count rate or radiation dose, and usually in the case of hand-held instruments, an audio device producing clicks.

The advantages are:

- Cheap, robust detector with a large variety of sizes and applications

- Large output signal from tube requiring minimal electronic processing for simple counting

- Can measure overall gamma dose when using energy compensated tube

The disadvantages are:

- Cannot measure energy of radiation - no spectrographic information

- Will not measure high radiation rates due to dead time

- Sustained high radiation levels will degrade fill gas

Guidance on Detector Type Usage

The UK Health and Safety Executive has issued a guidance note on the correct portable instrument for the application concerned . This covers all radiation instrument technologies and is useful in selecting the correct gaseous ionisation detector technology for a measurement application.

Everyday Use

Ionization-type smoke detectors are gaseous ionization detectors in widespread use. A small source of radioactive americium is placed so that it maintains a current between two plates that effectively form an ionisation chamber. If smoke gets between the plates where ionization is taking place, the ionized gas can be neutralized leading to a reduced current. The decrease in current triggers a fire alarm.

Scintillation Counter

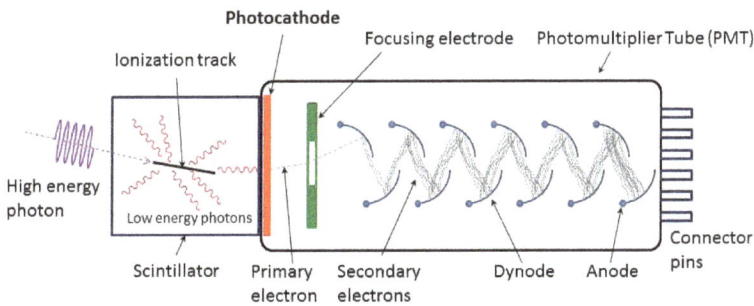

Schematic showing incident high energy photon hitting a scintillating crystal, triggering the release of low-energy photons which are then converted into photoelectrons and multiplied in the photomultiplier

A scintillation counter is an instrument for detecting and measuring ionizing radiation by using the excitation effect of incident radiation on a scintillator material, and detecting the resultant light pulses.

It consists of a scintillator which generates photons in response to incident radiation, a sensitive photomultiplier tube (PMT) which converts the light to an electrical signal and electronics to process this signal.

Scintillation counters are widely used in radiation protection, assay of radioactive materials and physics research because they can be made inexpensively yet with good quantum efficiency, and can measure both the intensity and the energy of incident radiation.

History

The modern electronic scintillation counter was invented in 1944 by Sir Samuel Curran whilst he was working on the Manhattan Project at the University of California at Berkeley. There was a requirement to measure the radiation from small quantities of uranium and his innovation was to use one of the newly-available highly sensitive photomultiplier tubes made by the Radio Corporation of America to accurately count the flashes of light from a scintillator subjected to radiation. This built upon the work of earlier researchers such as Antoine Henri Becquerel, who discovered radioactivity whilst working on the phosphorescence of uranium salts in 1896. Previously scintillation events had to be laboriously detected by eye using a spinthariscope which was a simple microscope to observe light flashes in the scintillator.

Operation

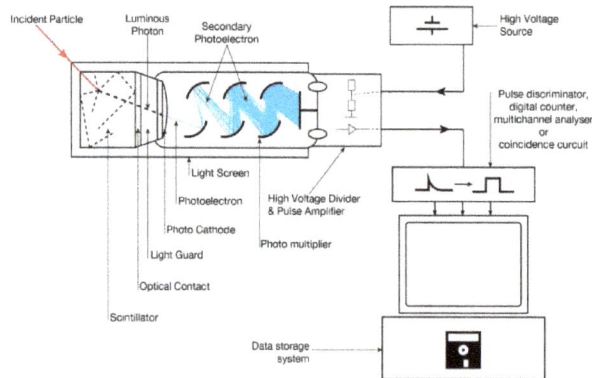

Apparatus with a scintillating crystal, photomultiplier, and data acquisition components.

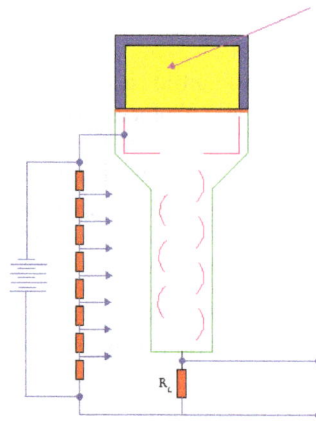

animation of radiation scintillation counter

When an ionizing particle passes into the scintillator material, atoms are ionized along a track. For charged particles the track is the path of the particle itself. For gamma rays (uncharged), their energy is converted to an energetic electron via either the photoelectric effect, Compton scattering or pair production. The chemistry of atomic de-excitation in the scintillator produces a multitude of low-energy photons, typically near the blue end of the visible spectrum. The number of such photons is in proportion to the amount of energy deposited by the ionizing particle. Some portion of these low-energy photons arrive at the photocathode of an attached photomultiplier tube. The photocathode emits at most one electron for each arriving photon by the photoelectric effect. This group of primary electrons is electrostatically accelerated and focused by an electrical potential so that they strike the first dynode of the tube. The impact of a single electron on the dynode releases a number of secondary electrons which are in turn accelerated to strike the second dynode. Each subsequent dynode impact releases further electrons, and so there is a current amplifying effect at each dynode stage. Each stage is at a higher potential than the previous to provide the accelerating field. The resultant output signal at the anode is in the form of a measurable pulse for each group of photons that arrived at the photocathode, and is passed to the processing electronics. The pulse carries information about the energy of the original incident radiation on the scintillator. The number of such pulses per unit time gives information about the intensity of the radiation. In some applications individual pulses are not counted, but rather only the average current at the anode is used as a measure of radiation intensity.

The scintillator must be shielded from all ambient light so that external photons do not swamp the ionization events caused by incident radiation. To achieve this a thin opaque foil, such as aluminized mylar, is often used, though it must have a low enough mass to minimize undue attenuation of the incident radiation being measured.

The article on the photomultiplier tube carries a detailed description of the tube's operation.

Detection Materials

The scintillator consists of a transparent crystal, usually a phosphor, plastic (usually containing anthracene) or organic liquid that fluoresces when struck by ionizing radiation.

Cesium iodide (CsI) in crystalline form is used as the scintillator for the detection of protons and alpha particles. Sodium iodide (NaI) containing a small amount of thallium is used as a scintillator for the detection of gamma waves and zinc sulfide (ZnS) is widely used as a detector of alpha particles. Zinc sulfide is the material Rutherford used to perform his scattering experiment. Lithium iodide (LiI) is used in neutron detectors.

Detector Efficiencies

Gamma

The quantum efficiency of a gamma-ray detector (per unit volume) depends upon the density of electrons in the detector, and certain scintillating materials, such as sodium iodide and bismuth germanate, achieve high electron densities as a result of the high atomic numbers of some of the elements of which they are composed. However, detectors based on semiconductors, notably hyperpure germanium, have better intrinsic energy resolution than scintillators, and are preferred where feasible for gamma-ray spectrometry.

Neutron

In the case of neutron detectors, high efficiency is gained through the use of scintillating materials rich in hydrogen that scatter neutrons efficiently. Liquid scintillation counters are an efficient and practical means of quantifying beta radiation.

Applications

Scintillation counters are used to measure radiation in a variety of applications including hand held radiation survey meters, personnel and environmental monitoring for radioactive contamination, medical imaging, radiometric assay, nuclear security and nuclear plant safety.

Several products have been introduced in the market utilising scintillation counters for detection of potentially dangerous gamma-emitting materials during transport. These include scintillation counters designed for freight terminals, border security, ports, weigh bridge applications, scrap metal yards and contamination monitoring of nuclear waste. There are variants of scintillation counters mounted on pick-up trucks and helicopters for rapid response in case of a security situation due to dirty bombs or radioactive waste. Hand-held units are also commonly used.

Guidance on Application Use

In the United Kingdom the HSE has issued a user guidance note on selecting the correct radiation measurement instrument for the application concerned . This covers all radiation instrument technologies, and is a useful comparative guide to the use of scintillation detectors.

Radiation Protection

Alpha and Beta Contamination

Hand-held large area alpha scintillation probe under calibration

Hand-held scintillation counter reading ambient gamma dose. The position of the internal detector is shown by the cross

Industrial radioactive contamination monitors, either hand-held for area or personal surveys or installed for personnel monitoring require a large detection area to ensure efficient and rapid coverage of monitored surfaces. For this the scintillation counter with a large area scintillator window and integrated photomultiplier tube is ideally suited and finds wide application in the field of radioactive contamination monitoring of personnel and the environment. Detectors are designed to have one or two scintillation materials, depending on the application. "Single phosphor" detectors are used for either alpha or beta, and "Dual phosphor" detectors are used to detect both.

A scintillator such as zinc sulphide is used for alpha particle detection, whilst plastic scintillators are used for beta detection. The resultant scintillation energies can be discriminated so that alpha and beta counts can be measured separately with the same detector. This technique is used in both hand-held and fixed monitoring equipment, and such instruments are relatively inexpensive compared with the gas proportional detector.

Gamma

Scintillation materials are used for ambient gamma dose measurement, though a different construction is used to detect contamination, as no thin window is required.

Scintillation Counter as a Spectrometer

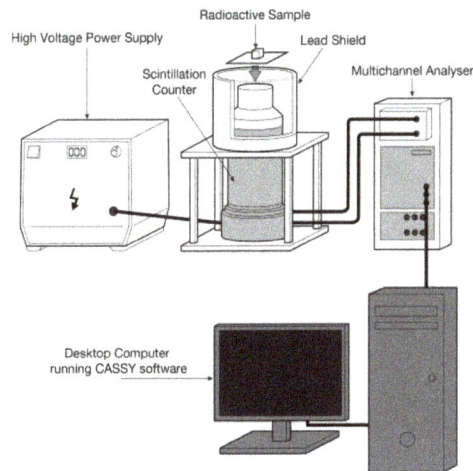

The experimental setup for determination of γ-radiation spectrum with a scintillation counter. A high voltage power supply is connected to the scintillation counter. The scintillation counter is connected to the Multichannel Analyser which sends information to the computer.

Scintillators often convert a single photon of high energy radiation into high number of lower-energy photons, where the number of photons per megaelectronvolt of input energy is fairly constant. By measuring the intensity of the flash (the number of the photons produced by the x-ray or gamma photon) it is therefore possible to discern the original photon's energy.

The spectrometer consists of a suitable scintillator crystal, a photomultiplier tube, and a circuit for measuring the height of the pulses produced by the photomultiplier. The pulses are counted and sorted by their height, producing a x-y plot of scintillator flash brightness vs number of the flashes, which approximates the energy spectrum of the incident radiation, with some additional artifacts. A monochromatic gamma radiation produces a photopeak at its energy. The detector also shows response at the lower energies, caused by Compton scattering, two smaller escape peaks at energies 0.511 and 1.022 MeV below the photopeak for the creation of electron-positron pairs when one or both annihilation photons escape, and a backscatter peak. Higher energies can be measured when two or more photons strike the detector almost simultaneously (pile-up, within the time resolution of the data acquisition chain), appearing as sum peaks with energies up to the value of two or more photopeaks added.

References

- International Atomic Energy Agency (2007). IAEA Safety Glossary: Terminology Used in Nuclear Safety and Radiation Protection (PDF). Vienna: IAEA. ISBN 92-0-100707-8.

- International Atomic Energy Agency (2010). Programmes and Systems for Source and Environmental Radiation Monitoring. Safety Reports Series No. 64. Vienna: IAEA. p. 234. ISBN 978-92-0-112409-8.

- Podgorsak, E. B., ed. (2005). Radiation Oncology Physics: A Handbook for Teachers and Students (PDF). Vienna: International Atomic Energy Agency. ISBN 92-0-107304-6. Retrieved 25 November 2012.

- Knoll, Glenn F (1999). Radiation detection and measurement (3rd ed.). New York: Wiley. ISBN 0-471-07338-5.

- For example, see Basseville and Nikiforov, Detection of Abrupt Changes: Theory and Application, Prentice-Hall (1993) ISBN 0-13-126780-9

6

Environmental Concerns and Challenges

Environmental concerns and challenges have drastically increased in the past few years. Some of the major concerns related to the environment are drug pollution, air pollution, greenhouse gas, acid rain and chemical waste. The pollution that is caused by pharmaceutical drugs by dumping their waste in the marine environment causes drug pollution as well as marine pollution. The topics discussed in the chapter are of great importance to broaden the existing knowledge on the environment.

Air Pollution

Air pollution is the introduction of particulates, biological molecules, or other harmful substances into Earth's atmosphere, causing diseases, allergies, death to humans, damage to other living organisms such as animals and food crops, or the natural or built environment. Air pollution may come from anthropogenic or natural sources.

Air pollution from a fossil-fuel power station

The atmosphere is a complex natural gaseous system that is essential to support life on planet Earth.

Indoor air pollution and urban air quality are listed as two of the world's worst toxic pollution problems in the 2008 Blacksmith Institute World's Worst Polluted Places report. According to the 2014 WHO report, air pollution in 2012 caused the deaths of around 7 million people worldwide, an estimate roughly matched by the International Energy Agency.

Pollutants

Carbon dioxide in Earth's atmosphere if *half* of global-warming emissions are *not* absorbed.
(NASA simulation; 9 November 2015)

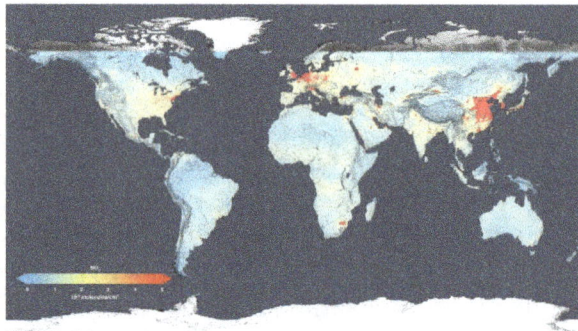

Nitrogen dioxide 2014 - global air quality levels
(released 14 December 2015).

An air pollutant is a substance in the air that can have adverse effects on humans and the ecosystem. The substance can be solid particles, liquid droplets, or gases. A pollutant can be of natural origin or man-made. Pollutants are classified as primary or secondary. Primary pollutants are usually produced from a process, such as ash from a volcanic eruption. Other examples include carbon monoxide gas from motor vehicle exhaust, or the sulfur dioxide released from factories. Secondary pollutants are not emitted directly. Rather, they form in the air when primary pollutants react or interact. Ground level ozone is a prominent example of a secondary pollutant. Some pollutants may be both primary and secondary: they are both emitted directly and formed from other primary pollutants.

Before flue-gas desulfurization was installed, the emissions from this power plant in New Mexico contained excessive amounts of sulfur dioxide.

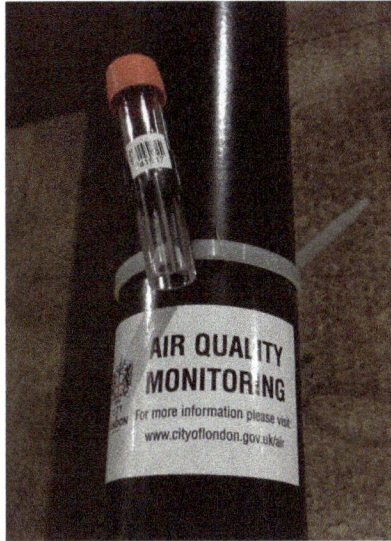

Nitrogen dioxide diffusion tube for air quality monitoring. Positioned in London City.

Major primary pollutants produced by human activity include:

- Sulfur oxides (SO_x) - particularly sulfur dioxide, a chemical compound with the formula SO_2. SO_2 is produced by volcanoes and in various industrial processes. Coal and petroleum often contain sulfur compounds, and their combustion generates sulfur dioxide. Further oxidation of SO_2, usually in the presence of a catalyst such as NO_2, forms H_2SO_4, and thus acid rain. This is one of the causes for concern over the environmental impact of the use of these fuels as power sources.

- Nitrogen oxides (NO_x) - Nitrogen oxides, particularly nitrogen dioxide, are expelled from high temperature combustion, and are also produced during thunderstorms by electric discharge. They can be seen as a brown haze dome above or a plume downwind of cities. Nitrogen dioxide is a chemical compound with the formula NO_2. It is one of several nitrogen oxides. One of the most prominent air pollutants, this reddish-brown toxic gas has a characteristic sharp, biting odor.

- Carbon monoxide (CO) - CO is a colorless, odorless, toxic yet non-irritating gas. It is a product by incomplete combustion of fuel such as natural gas, coal or wood. Vehicular exhaust is a major source of carbon monoxide.

- Volatile organic compounds (VOC) - VOCs are a well-known outdoor air pollutant. They are categorized as either methane (CH_4) or non-methane (NMVOCs). Methane is an extremely efficient greenhouse gas which contributes to enhanced global warming. Other hydrocarbon VOCs are also significant greenhouse gases because of their role in creating ozone and prolonging the life of methane in the atmosphere. This effect varies depending on local air quality. The aromatic

NMVOCs benzene, toluene and xylene are suspected carcinogens and may lead to leukemia with prolonged exposure. 1,3-butadiene is another dangerous compound often associated with industrial use.

- Particulates, alternatively referred to as particulate matter (PM), atmospheric particulate matter, or fine particles, are tiny particles of solid or liquid suspended in a gas. In contrast, aerosol refers to combined particles and gas. Some particulates occur naturally, originating from volcanoes, dust storms, forest and grassland fires, living vegetation, and sea spray. Human activities, such as the burning of fossil fuels in vehicles, power plants and various industrial processes also generate significant amounts of aerosols. Averaged worldwide, anthropogenic aerosols—those made by human activities—currently account for approximately 10 percent of our atmosphere. Increased levels of fine particles in the air are linked to health hazards such as heart disease, altered lung function and lung cancer.

- Persistent free radicals connected to airborne fine particles are linked to cardiopulmonary disease.

- Toxic metals, such as lead and mercury, especially their compounds.

- Chlorofluorocarbons (CFCs) - harmful to the ozone layer; emitted from products are currently banned from use. These are gases which are released from air conditioners, refrigerators, aerosol sprays, etc. CFC's on being released into the air rises to stratosphere. Here they come in contact with other gases and damage the ozone layer. This allows harmful ultraviolet rays to reach the earth's surface. This can lead to skin cancer, disease to eye and can even cause damage to plants.

- Ammonia (NH_3) - emitted from agricultural processes. Ammonia is a compound with the formula NH_3. It is normally encountered as a gas with a characteristic pungent odor. Ammonia contributes significantly to the nutritional needs of terrestrial organisms by serving as a precursor to foodstuffs and fertilizers. Ammonia, either directly or indirectly, is also a building block for the synthesis of many pharmaceuticals. Although in wide use, ammonia is both caustic and hazardous. In the atmosphere, ammonia reacts with oxides of nitrogen and sulfur to form secondary particles.

- Odours — such as from garbage, sewage, and industrial processes

- Radioactive pollutants - produced by nuclear explosions, nuclear events, war explosives, and natural processes such as the radioactive decay of radon.

Secondary pollutants include:

- Particulates created from gaseous primary pollutants and compounds in photochemical smog. Smog is a kind of air pollution. Classic smog results from large amounts of coal burning in an area caused by a mixture of smoke and sulfur dioxide. Modern smog does not usually come from coal but from vehicular and industrial emissions that are acted on in the atmosphere by ultraviolet light from the sun to form secondary pollutants that also combine with the primary emissions to form photochemical smog.

- Ground level ozone (O_3) formed from NO_x and VOCs. Ozone (O_3) is a key constituent of the troposphere. It is also an important constituent of certain regions of the stratosphere commonly known as the Ozone layer. Photochemical and chemical reactions involving it drive many of the chemical processes that occur in the atmosphere by day and by night. At abnormally high concentrations brought about by human activities (largely the combustion of fossil fuel), it is a pollutant, and a constituent of smog.

- Peroxyacetyl nitrate (PAN) - similarly formed from NO_x and VOCs.

Minor air pollutants include:

- A large number of minor hazardous air pollutants. Some of these are regulated in USA under the Clean Air Act and in Europe under the Air Framework Directive

- A variety of persistent organic pollutants, which can attach to particulates

Persistent organic pollutants (POPs) are organic compounds that are resistant to environmental degradation through chemical, biological, and photolytic processes. Because of this, they have been observed to persist in the environment, to be capable of long-range transport, bioaccumulate in human and animal tissue, biomagnify in food chains, and to have potentially significant impacts on human health and the environment.

Sources

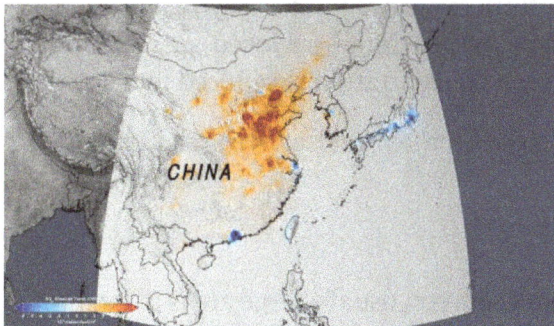

This video provides an overview of a NASA study on the human fingerprint on global air quality.

There are various locations, activities or factors which are responsible for releasing pollutants into the atmosphere. These sources can be classified into two major categories.

Anthropogenic (man-made) sources:

Controlled burning of a field outside of Statesboro, Georgia in preparation for spring planting.

These are mostly related to the burning of multiple types of fuel.

- Stationary sources include smoke stacks of power plants, manufacturing facilities (factories) and waste incinerators, as well as furnaces and other types of fuel-burning heating devices. In developing and poor countries, traditional biomass burning is the major source of air pollutants; traditional biomass includes wood, crop waste and dung.

- Mobile sources include motor vehicles, marine vessels, and aircraft.

- Controlled burn practices in agriculture and forest management. Controlled or prescribed burning is a technique sometimes used in forest management, farming, prairie restoration or greenhouse gas abatement. Fire is a natural part of both forest and grassland ecology and controlled fire can be a tool for foresters. Controlled burning stimulates the germination of some desirable forest trees, thus renewing the forest.

- Fumes from paint, hair spray, varnish, aerosol sprays and other solvents

- Waste deposition in landfills, which generate methane. Methane is highly flammable and may form explosive mixtures with air. Methane is also an asphyxiant and may displace oxygen in an enclosed space. Asphyxia or suffocation may result if the oxygen concentration is reduced to below 19.5% by displacement.

- Military resources, such as nuclear weapons, toxic gases, germ warfare and rocketry

Natural sources:

- Dust from natural sources, usually large areas of land with little or no vegetation

- Methane, emitted by the digestion of food by animals, for example cattle

- Radon gas from radioactive decay within the Earth's crust. Radon is a colorless, odorless, naturally occurring, radioactive noble gas that is formed from the decay of radium. It is considered to be a health hazard. Radon gas from natural sources can accumulate in buildings, especially in confined areas such as the basement and it is the second most frequent cause of lung cancer, after cigarette smoking.

Dust storm approaching Stratford, Texas.

- Smoke and carbon monoxide from wildfires

- Vegetation, in some regions, emits environmentally significant amounts of Volatile organic compounds (VOCs) on warmer days. These VOCs react with primary anthropogenic pollutants—specifically, NO_x, SO_2, and anthropogenic organic carbon compounds — to produce a seasonal haze of secondary pollutants. Black gum, poplar, oak and willow are some examples of vegetation that can produce abundant VOCs. The VOC production from these species result in ozone levels up to eight times higher than the low-impact tree species.

- Volcanic activity, which produces sulfur, chlorine, and ash particulates

Emission Factors

Beijing air on a 2005-day after rain (left) and a smoggy day (right)

Air pollutant emission factors are reported representative values that attempt to relate the quantity of a pollutant released to the ambient air with an activity associated with the release of that pollutant. These factors are usually expressed as the weight of pollutant divided by a unit weight, volume, distance, or duration of the activity emitting the pollutant (e.g., kilograms of particulate emitted per tonne of coal burned). Such factors facilitate estimation of emissions from various sources of air pollution. In most cases, these factors are simply averages of all available data of acceptable quality, and are generally assumed to be representative of long-term averages.

There are 12 compounds in the list of persistent organic pollutants. Dioxins and furans are two of them and intentionally created by combustion of organics, like open burning of plastics. These compounds are also endocrine disruptors and can mutate the human genes.

The United States Environmental Protection Agency has published a compilation of air pollutant emission factors for a multitude of industrial sources. The United Kingdom, Australia, Canada and many other countries have published similar compilations, as well as the European Environment Agency.

Exposure

Air pollution risk is a function of the hazard of the pollutant and the exposure to that pollutant. Air pollution exposure can be expressed for an individual, for certain groups (e.g. neighborhoods or children living in a country), or for entire populations. For example, one may want to calculate the exposure to a hazardous air pollutant for a geographic area, which includes the various microenvironments and age groups. This can be calculated as an inhalation exposure. This would account for daily exposure in various settings (e.g. different indoor micro-environments and outdoor locations). The exposure needs to include different age and other demographic groups, especially infants, children, pregnant women and other sensitive subpopulations. The exposure to an air pollutant must integrate the concentrations of the air pollutant with respect to the time spent in each setting and the respective inhalation rates for each subgroup for each specific time that the subgroup is in the setting and engaged in particular activities (playing, cooking, reading, working, etc.). For example, a small child's inhalation rate will be less than that of an adult. A child engaged in vigorous exercise will have a higher respiration rate than the same child in a sedentary activity. The daily exposure, then, needs to reflect the time spent in each micro-environmental setting and the type of activities in these settings. The air pollutant concentration in each microactivity/microenvironmental setting is summed to indicate the exposure.

Indoor Air Quality (IAQ)

A lack of ventilation indoors concentrates air pollution where people often spend the majority of their time. Radon (Rn) gas, a carcinogen, is exuded from the Earth in certain locations and trapped inside houses. Building materials including carpeting and plywood emit formaldehyde (H_2CO) gas. Paint and solvents give off volatile organic compounds (VOCs) as they dry. Lead paint can degenerate into dust and be inhaled. Intentional air pollution is introduced with the use of air fresheners, incense, and other scented items.

Controlled wood fires in stoves and fireplaces can add significant amounts of smoke particulates into the air, inside and out. Indoor pollution fatalities may be caused by using pesticides and other chemical sprays indoors without proper ventilation.

Air quality monitoring, New Delhi, India.

Carbon monoxide poisoning and fatalities are often caused by faulty vents and chimneys, or by the burning of charcoal indoors or in a confined space, such as a tent. Chronic carbon monoxide poisoning can result even from poorly-adjusted pilot lights. Traps are built into all domestic plumbing to keep sewer gas and hydrogen sulfide, out of interiors. Clothing emits tetrachloroethylene, or other dry cleaning fluids, for days after dry cleaning.

Though its use has now been banned in many countries, the extensive use of asbestos in industrial and domestic environments in the past has left a potentially very dangerous material in many localities. Asbestosis is a chronic inflammatory medical condition affecting the tissue of the lungs. It occurs after long-term, heavy exposure to asbestos from asbestos-containing materials in structures. Sufferers have severe dyspnea (shortness of breath) and are at an increased risk regarding several different types of lung cancer. As clear explanations are not always stressed in non-technical literature, care should be taken to distinguish between several forms of relevant diseases. According to the World Health Organisation (WHO), these may defined as; asbestosis, *lung cancer*, and *Peritoneal Mesothelioma* (generally a very rare form of cancer, when more widespread it is almost always associated with prolonged exposure to asbestos).

Biological sources of air pollution are also found indoors, as gases and airborne particulates. Pets produce dander, people produce dust from minute skin flakes and decomposed hair, dust mites in bedding, carpeting and furniture produce enzymes and micrometre-sized fecal droppings, inhabitants emit methane, mold forms on walls and generates mycotoxins and spores, air conditioning systems can incubate Legionnaires' disease and mold, and houseplants, soil and surrounding gardens can produce pollen, dust, and mold. Indoors, the lack of air circulation allows these airborne pollutants to accumulate more than they would otherwise occur in nature.

Health Effects

Air pollution is a significant risk factor for a number of pollution-related diseases and health conditions including respiratory infections, heart disease, COPD, stroke and lung cancer. The health effects caused by air pollution may include difficulty in breathing, wheezing, coughing, asthma and worsening of existing respiratory and cardiac conditions. These effects can result in increased medication use, increased doctor or emergency room visits, more hospital admissions and premature death. The human health effects of poor air quality are far reaching, but principally affect the body's respiratory system and the cardiovascular system. Individual reactions to air pollutants depend on the type of pollutant a person is exposed to, the degree of exposure, and the individual's health status and genetics. The most common sources of air pollution include particulates, ozone, nitrogen dioxide, and sulphur dioxide. Children aged less than five years that live in developing countries are the most vulnerable population in terms of total deaths attributable to indoor and outdoor air pollution.

Mortality

The World Health Organization estimated in 2014 that every year air pollution causes the premature death of some 7 million people worldwide. India has the highest death rate due to air pollution. India also has more deaths from asthma than any other nation according to the World Health Organization. In December 2013 air pollution was estimated to kill 500,000 people in China each year. There is a positive correlation between pneumonia-related deaths and air pollution from motor vehicle emissions.

Annual premature European deaths caused by air pollution are estimated at 430,000. An important cause of these deaths is nitrogen dioxide and other nitrogen oxides (NOx) emitted by road vehicles. Across the European Union, air pollution is estimated to reduce life expectancy by almost nine months. Causes of deaths include strokes, heart disease, COPD, lung cancer, and lung infections.

The US EPA estimates that a proposed set of changes in diesel engine technology (*Tier 2*) could result in 12,000 fewer *premature mortalities*, 15,000 fewer heart attacks, 6,000 fewer emergency room visits by children with asthma, and 8,900 fewer respiratory-related hospital admissions each year in the United States.

The US EPA has estimated that limiting ground-level ozone concentration to 65 parts per billion, would avert 1,700 to 5,100 premature deaths nationwide in 2020 compared with the 75-ppb standard. The agency projected the more protective standard would also prevent an additional 26,000 cases of aggravated asthma, and more than a million cases of missed work or school. Following this assessment, the EPA acted to protect public health by lowering the National Ambient Air Quality Standards (NAAQS) for ground-level ozone to 70 parts per billion (ppb).

A new economic study of the health impacts and associated costs of air pollution in the

Los Angeles Basin and San Joaquin Valley of Southern California shows that more than 3,800 people die prematurely (approximately 14 years earlier than normal) each year because air pollution levels violate federal standards. The number of annual premature deaths is considerably higher than the fatalities related to auto collisions in the same area, which average fewer than 2,000 per year.

Diesel exhaust (DE) is a major contributor to combustion-derived particulate matter air pollution. In several human experimental studies, using a well-validated exposure chamber setup, DE has been linked to acute vascular dysfunction and increased thrombus formation.

The mechanisms linking air pollution to increased cardiovascular mortality are uncertain, but probably include pulmonary and systemic inflammation.

Cardiovascular Disease

A 2007 review of evidence found ambient air pollution exposure is a risk factor correlating with increased total mortality from cardiovascular events (range: 12% to 14% per 10 microg/m^3 increase).

Air pollution is also emerging as a risk factor for stroke, particularly in developing countries where pollutant levels are highest. A 2007 study found that in women, air pollution is not associated with hemorrhagic but with ischemic stroke. Air pollution was also found to be associated with increased incidence and mortality from coronary stroke in a cohort study in 2011. Associations are believed to be causal and effects may be mediated by vasoconstriction, low-grade inflammation and atherosclerosis Other mechanisms such as autonomic nervous system imbalance have also been suggested.

Lung Disease

Chronic obstructive pulmonary disease (COPD) includes diseases such as chronic bronchitis and emphysema.

Research has demonstrated increased risk of developing asthma and COPD from increased exposure to traffic-related air pollution. Additionally, air pollution has been associated with increased hospitalization and mortality from asthma and COPD.

A study conducted in 1960-1961 in the wake of the Great Smog of 1952 compared 293 London residents with 477 residents of Gloucester, Peterborough, and Norwich, three towns with low reported death rates from chronic bronchitis. All subjects were male postal truck drivers aged 40 to 59. Compared to the subjects from the outlying towns, the London subjects exhibited more severe respiratory symptoms (including cough, phlegm, and dyspnea), reduced lung function (FEV$_1$ and peak flow rate), and increased sputum production and purulence. The differences were more pronounced for subjects aged 50 to 59. The study controlled for age and smoking habits, so concluded that air pollution was the most likely cause of the observed differences.

It is believed that much like cystic fibrosis, by living in a more urban environment serious health hazards become more apparent. Studies have shown that in urban areas patients suffer mucus hypersecretion, lower levels of lung function, and more self-diagnosis of chronic bronchitis and emphysema.

Cancer

Cancer mainly the result of environmental factors.

A review of evidence regarding whether ambient air pollution exposure is a risk factor for cancer in 2007 found solid data to conclude that long-term exposure to PM2.5 (fine particulates) increases the overall risk of non-accidental mortality by 6% per a 10 microg/m³ increase. Exposure to PM2.5 was also associated with an increased risk of mortality from lung cancer (range: 15% to 21% per 10 microg/m³ increase) and total cardiovascular mortality (range: 12% to 14% per a 10 microg/m³ increase). The review further noted that living close to busy traffic appears to be associated with elevated risks of these three outcomes --- increase in lung cancer deaths, cardiovascular deaths, and overall non-accidental deaths. The reviewers also found suggestive evidence that exposure to PM2.5 is positively associated with mortality from coronary heart diseases and exposure to SO_2 increases mortality from lung cancer, but the data was insufficient to provide solid conclusions. Another investigation showed that higher activity level increases deposition fraction of aerosol particles in human lung and recommended avoiding heavy activities like running in outdoor space at polluted areas.

In 2011, a large Danish epidemiological study found an increased risk of lung cancer for patients who lived in areas with high nitrogen oxide concentrations. In this study, the association was higher for non-smokers than smokers. An additional Danish study, also in 2011, likewise noted evidence of possible associations between air pollution and other forms of cancer, including cervical cancer and brain cancer.

In December 2015, medical scientists reported that cancer is overwhelmingly a result of environmental factors, and not largely down to bad luck. Maintaining a healthy weight, eating a healthy diet, minimizing alcohol and eliminating smoking reduces the risk of developing the disease, according to the researchers.

Children

In the United States, despite the passage of the Clean Air Act in 1970, in 2002 at least 146 million Americans were living in non-attainment areas—regions in which the concentration of certain air pollutants exceeded federal standards. These dangerous pollutants are known as the criteria pollutants, and include ozone, particulate matter, sulfur dioxide, nitrogen dioxide, carbon monoxide, and lead. Protective measures to ensure children's health are being taken in cities such as New Delhi, India where buses now use compressed natural gas to help eliminate the "pea-soup" smog. A recent study in Europe has found that exposure to ultrafine particles can increase blood pressure in children.

"Clean" Areas

Even in the areas with relatively low levels of air pollution, public health effects can be significant and costly, since a large number of people breathe in such pollutants. A 2005 scientific study for the British Columbia Lung Association showed that a small improvement in air quality (1% reduction of ambient PM2.5 and ozone concentrations) would produce $29 million in annual savings in the Metro Vancouver region in 2010. This finding is based on health valuation of lethal (death) and sub-lethal (illness) affects.

Central Nervous System

Data is accumulating that air pollution exposure also affects the central nervous system.

In a June 2014 study conducted by researchers at the University of Rochester Medical Center, published in the journal Environmental Health Perspectives, it was discovered that early exposure to air pollution causes the same damaging changes in the brain as autism and schizophrenia. The study also shows that air pollution also affected short-term memory, learning ability, and impulsivity. Lead researcher Professor Deborah Cory-Slechta said that "When we looked closely at the ventricles, we could see that the white matter that normally surrounds them hadn't fully developed. It appears that inflammation had damaged those brain cells and prevented that region of the brain from developing, and the ventricles simply expanded to fill the space. Our findings add to the growing body of evidence that air pollution may play a role in autism, as well as in other neurodevelopmental disorders." Air pollution has a more significant negative effect on males than on females.

In 2015, experimental studies reported the detection of significant episodic (situational) cognitive impairment from impurities in indoor air breathed by test subjects who were not informed about changes in the air quality. Researchers at the Harvard University and SUNY Upstate Medical University and Syracuse University measured the cognitive performance of 24 participants in three different controlled laboratory atmospheres that simulated those found in "conventional" and "green" buildings, as well as green buildings with enhanced ventilation. Performance was evaluated objectively using the widely used Strategic Management Simulation software simulation tool, which is a well-validated assessment test for executive decision-making in an unconstrained situation allowing initiative and improvisation. Significant deficits were observed in the performance scores achieved in increasing concentrations of either volatile organic compounds (VOCs) or carbon dioxide, while keeping other factors constant. The highest impurity levels reached are not uncommon in some classroom or office environments.

Agricultural Effects

In India in 2014, it was reported that air pollution by black carbon and ground level ozone had cut crop yields in the most affected areas by almost half in 2010 when compared to 1980 levels.

Economic Effects

Air pollution costs the world economy $5 trillion per year as a result of productivity losses and degraded quality of life, according to a joint study by the World Bank and the Institute for Health Metrics and Evaluation (IHME) at the University of Washington These productivity losses are caused by deaths due to diseases caused by air pollution. One out of ten deaths in 2013 was caused by diseases associated with air pollution and the problem is getting worse. The problem is even more acute in the developing world. "Children under age 5 in lower-income countries are more than 60 times as likely to die from exposure to air pollution as children in high-income countries." The report states that additional economic losses caused by air pollution, including health costs and the adverse effect on agricultural and other productivity were not calculated in the report, and thus the actual costs to the world economy are far higher than $5 trillion.

Historical Disasters

The world's worst short-term civilian pollution crisis was the 1984 Bhopal Disaster in India. Leaked industrial vapours from the Union Carbide factory, belonging to Union Carbide, Inc., U.S.A. (later bought by Dow Chemical Company), killed at least 3787 people and injured anywhere from 150,000 to 600,000. The United Kingdom suffered its worst air pollution event when the December 4 Great Smog of 1952 formed over London. In six days more than 4,000 died and more recent estimates put the figure at nearer 12,000. An accidental leak of anthrax spores from a biological warfare laboratory in the former USSR in 1979 near Sverdlovsk is believed to have caused at least 64 deaths. The worst single incident of air pollution to occur in the US occurred in Donora, Pennsylvania in late October, 1948, when 20 people died and over 7,000 were injured.

Alternatives to Pollution

There are now practical alternatives to the three principal causes of air pollution.

- Combustion of fossil fuels for space heating can be replaced by using ground source heat pumps and seasonal thermal energy storage.

- Electric power generation from burning fossil fuels can be replaced by power generation from nuclear and renewables.

- Motor vehicles driven by fossil fuels, a key factor in urban air pollution, can be replaced by electric vehicles.

Reduction Efforts

There are various air pollution control technologies and strategies available to reduce air pollution. At its most basic level, land-use planning is likely to involve zoning and transport infrastructure planning. In most developed countries, land-use planning is an important part of social policy, ensuring that land is used efficiently for the benefit of the wider economy and population, as well as to protect the environment.

Because a large share of air pollution is caused by combustion of fossil fuels such as coal and oil, the reduction of these fuels can reduce air pollution drastically. Most effective is the switch to clean power sources such as wind power, solar power, hydro power which don't cause air pollution. Efforts to reduce pollution from mobile sources includes primary regulation (many developing countries have permissive regulations), expanding regulation to new sources (such as cruise and transport ships, farm equipment, and small gas-powered equipment such as string trimmers, chainsaws, and snowmobiles), increased fuel efficiency (such as through the use of hybrid vehicles), conversion to cleaner fuels or conversion to electric vehicles.

Titanium dioxide has been researched for its ability to reduce air pollution. Ultraviolet light will release free electrons from material, thereby creating free radicals, which break up VOCs and NOx gases. One form is superhydrophilic.

In 2014, Prof. Tony Ryan and Prof. Simon Armitage of University of Sheffield prepared a 10 meter by 20 meter-sized poster coated with microscopic, pollution-eating nanoparticles of titanium dioxide. Placed on a building, this giant poster can absorb the toxic emission from around 20 cars each day.

A very effective means to reduce air pollution is the transition to renewable energy. According to a study published in Energy and Environmental Science in 2015 the switch to 100% renewable energy in the United States would eliminate about 62,000 premature mortalities per year and about 42,000 in 2050, if no biomass were used. This would save about $600 billion in health costs a year due to reduced air pollution in 2050, or about 3.6% of the 2014 U.S. gross domestic product.

Control Devices

The following items are commonly used as pollution control devices in industry and transportation. They can either destroy contaminants or remove them from an exhaust stream before it is emitted into the atmosphere.

- Particulate control

    o Mechanical collectors (dust cyclones, multicyclones)

    o Electrostatic precipitators An electrostatic precipitator (ESP), or electrostatic air cleaner is a particulate collection device that removes particles from a flowing gas (such as air), using the force of an induced electrostatic charge. Electrostatic precipitators are highly efficient filtration devices that minimally impede the flow of gases through the device, and can easily remove fine particulates such as dust and smoke from the air stream.

    o Baghouses Designed to handle heavy dust loads, a dust collector consists of a blower, dust filter, a filter-cleaning system, and a dust receptacle or dust removal system (distinguished from air cleaners which utilize disposable filters to remove the dust).

    o Particulate scrubbers Wet scrubber is a form of pollution control technology. The term describes a variety of devices that use pollutants from a furnace flue gas or from other gas streams. In a wet scrubber, the polluted gas stream is brought into contact with the scrubbing liquid, by spraying it with the liquid, by forcing it through a pool of liquid, or by some other contact method, so as to remove the pollutants.

- Scrubbers

    o Baffle spray scrubber

    o Cyclonic spray scrubber

    o Ejector venturi scrubber

    o Mechanically aided scrubber

    o Spray tower

    o Wet scrubber

- NOx control

    o   Low NOx burners

    o   Selective catalytic reduction (SCR)

    o   Selective non-catalytic reduction (SNCR)

    o   NOx scrubbers

    o   Exhaust gas recirculation

    o   Catalytic converter (also for VOC control)

- VOC abatement

    o   Adsorption systems, using activated carbon, such as Fluidized Bed Concentrator

    o   Flares

    o   Thermal oxidizers

    o   Catalytic converters

    o   Biofilters

    o   Absorption (scrubbing)

    o   Cryogenic condensers

    o   Vapor recovery systems

- Acid Gas/SO_2 control

    o   Wet scrubbers

    o   Dry scrubbers

    o   Flue-gas desulfurization

- Mercury control

    o   Sorbent Injection Technology

    o   Electro-Catalytic Oxidation (ECO)

    o   K-Fuel

- Dioxin and furan control

- Miscellaneous associated equipment

      o   Source capturing systems

      o   Continuous emissions monitoring systems (CEMS)

Regulations

Smog in Cairo

In general, there are two types of air quality standards. The first class of standards (such as the U.S. National Ambient Air Quality Standards and E.U. Air Quality Directive) set maximum atmospheric concentrations for specific pollutants. Environmental agencies enact regulations which are intended to result in attainment of these target levels. The second class (such as the North American Air Quality Index) take the form of a scale with various thresholds, which is used to communicate to the public the relative risk of outdoor activity. The scale may or may not distinguish between different pollutants.

Canada

In Canada, air pollution and associated health risks are measured with the Air Quality Health Index or (AQHI). It is a health protection tool used to make decisions to reduce short-term exposure to air pollution by adjusting activity levels during increased levels of air pollution.

The Air Quality Health Index or "AQHI" is a federal program jointly coordinated by Health Canada and Environment Canada. However, the AQHI program would not be possible without the commitment and support of the provinces, municipalities and NGOs. From air quality monitoring to health risk communication and community engagement, local partners are responsible for the vast majority of work related to AQHI implementation. The AQHI provides a number from 1 to 10+ to indicate the level of health risk associated with local air quality. Occasionally, when the amount of air pollution is abnormally high, the number may exceed 10. The AQHI provides a local air quality current value as well as a local air quality maximums forecast for today, tonight and tomorrow and provides associated health advice.

| 1 | 2 | 3 | 4 | 5 | 6 | 7 | 8 | 9 | 10 | + |

| Risk: | Low (1-3) | Moderate (4-6) | High (7-10) | Very high (above 10) |

As it is now known that even low levels of air pollution can trigger discomfort for the sensitive population, the index has been developed as a continuum: The higher the number, the greater the health risk and need to take precautions. The index describes the level of health risk associated with this number as 'low', 'moderate', 'high' or 'very high', and suggests steps that can be taken to reduce exposure.

Health Risk	Air Quality Health Index	Health Messages	
		At Risk population	General Population
Low	1-3	Enjoy your usual outdoor activities.	Ideal air quality for outdoor activities
Moderate	4-6	Consider reducing or rescheduling strenuous activities outdoors if you are experiencing symptoms.	No need to modify your usual outdoor activities unless you experience symptoms such as coughing and throat irritation.
High	7-10	Reduce or reschedule strenuous activities outdoors. Children and the elderly should also take it easy.	Consider reducing or rescheduling strenuous activities outdoors if you experience symptoms such as coughing and throat irritation.
Very high	Above 10	Avoid strenuous activities outdoors. Children and the elderly should also avoid outdoor physical exertion and should stay indoors.	Reduce or reschedule strenuous activities outdoors, especially if you experience symptoms such as coughing and throat irritation.

The measurement is based on the observed relationship of Nitrogen Dioxide (NO_2), ground-level Ozone (O_3) and particulates ($PM_{2.5}$) with mortality, from an analysis of several Canadian cities. Significantly, all three of these pollutants can pose health risks, even at low levels of exposure, especially among those with pre-existing health problems.

When developing the AQHI, Health Canada's original analysis of health effects included five major air pollutants: particulates, ozone, and nitrogen dioxide (NO2), as well as sulfur dioxide (SO_2), and carbon monoxide (CO). The latter two pollutants provided little information in predicting health effects and were removed from the AQHI formulation.

The AQHI does not measure the effects of odour, pollen, dust, heat or humidity.

Germany

TA Luft is the German air quality regulation.

Hotspots

Air pollution hotspots are areas where air pollution emissions expose individuals to increased negative health effects. They are particularly common in highly populated, urban areas, where there may be a combination of stationary sources (e.g. industrial facilities) and mobile sources (e.g. cars and trucks) of pollution. Emissions from these sources can cause respiratory disease, childhood asthma, cancer, and other health problems. Fine particulate matter such as diesel soot, which contributes to more than 3.2 million premature deaths around the world each year, is a significant problem. It is very small and can lodge itself within the lungs and enter the bloodstream. Diesel soot is concentrated in densely populated areas, and one in six people in the U.S. live near a diesel pollution hot spot.

While air pollution hotspots affect a variety of populations, some groups are more likely to be located in hotspots. Previous studies have shown disparities in exposure to pollution by race and/or income. Hazardous land uses (toxic storage and disposal facilities, manufacturing facilities, major roadways) tend to be located where property values and income levels are low. Low socioeconomic status can be a proxy for other kinds of social vulnerability, including race, a lack of ability to influence regulation and a lack of ability to move to neighborhoods with less environmental pollution. These communities bear a disproportionate burden of environmental pollution and are more likely to face health risks such as cancer or asthma.

Studies show that patterns in race and income disparities not only indicate a higher exposure to pollution but also higher risk of adverse health outcomes. Communities characterized by low socioeconomic status and racial minorities can be more vulnerable to cumulative adverse health impacts resulting from elevated exposure to pollutants than more privileged communities. Blacks and Latinos generally face more pollution than whites and Asians, and low-income communities bear a higher burden of risk than affluent ones. Racial discrepancies are particularly distinct in suburban areas of the US South and metropolitan areas of the US West. Residents in public housing, who are generally low-income and cannot move to healthier neighborhoods, are highly affected by nearby refineries and chemical plants.

Cities

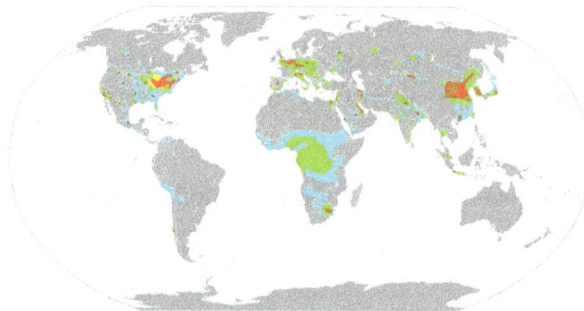

Nitrogen dioxide concentrations as measured from satellite 2002-2004

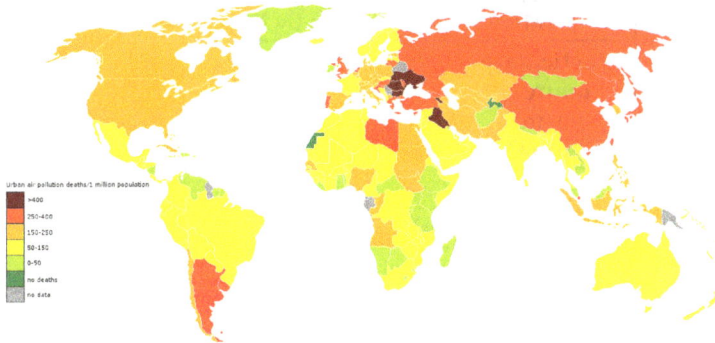

Deaths from air pollution in 2004

Air pollution is usually concentrated in densely populated metropolitan areas, especially in developing countries where environmental regulations are relatively lax or nonexistent. However, even populated areas in developed countries attain unhealthy levels of pollution, with Los Angeles and Rome being two examples. Between 2002 and 2011 the incidence of lung cancer in Beijing near doubled. While smoking remains the leading cause of lung cancer in China, the number of smokers is falling while lung cancer rates are rising. Another project focusing on the effects on pollution in vegetation has been researched by the local university in Sheffield, UK.

National-scale Air Toxics Assessments 1995-2005

The national-scale air toxics assessment(NATA) is an evaluation of air toxics by the U.S. EPA. EPA has furnished four assessments that characterize nationwide chronic cancer risk estimates and noncancer hazards from inhaling air toxics. The lates was from 2005, and made publicly available in early 2011.

"EPA developed the NATA as a state-of-the-science screening tool for State/Local/ Tribal Agencies to prioritize pollutants, emission sources and locations of interest for further study, in order to gain a better understanding of the risks. NATA assessments do not incorporate refined information about emission sources, but rather, use general information about sources to develop estimates of risks which are more likely to overestimate impacts than underestimate them. NATA provides estimates of the risk of cancer and other serious health effects from breathing (inhaling) air toxics in order to inform both national and more localized efforts to identify and prioritize air toxics, emission source types and locations which are of greatest potential concern in terms of contributing to population risk. This in turn helps air pollution experts focus limited analytical resources on areas and or populations where the potential for health risks are highest. Assessments include estimates of cancer and non-cancer health effects based on chronic exposure from outdoor sources, including assessments of non-cancer health effects for Diesel Particulate Matter. Assessments provide a snapshot of the outdoor air quality and the risks to human health that would result if air toxic emissions levels remained unchanged."

Most polluted cities by PM	
Particulate matter, μg/m³ (2004)	City
168	Cairo, Egypt
150	Delhi, India
128	Kolkata, India (Calcutta)
125	Tianjin, China
123	Chongqing, China
109	Kanpur, India
109	Lucknow, India
104	Jakarta, Indonesia
101	Shenyang, China

Governing Urban Air Pollution

In Europe, Council Directive 96/62/EC on ambient air quality assessment and management provides a common strategy against which member states can "set objectives for ambient air quality in order to avoid, prevent or reduce harmful effects on human health and the environment . . . and improve air quality where it is unsatisfactory".

On 25 July 2008 in the case Dieter Janecek v Freistaat Bayern CURIA, the European Court of Justice ruled that under this directive citizens have the right to require national authorities to implement a short term action plan that aims to maintain or achieve compliance to air quality limit values.

This important case law appears to confirm the role of the EC as centralised regulator to European nation-states as regards air pollution control. It places a supranational legal obligation on the UK to protect its citizens from dangerous levels of air pollution, furthermore superseding national interests with those of the citizen.

In 2010, the European Commission (EC) threatened the UK with legal action against the successive breaching of PM10 limit values. The UK government has identified that if fines are imposed, they could cost the nation upwards of £300 million per year.

In March 2011, the Greater London Built-up Area remains the only UK region in breach of the EC's limit values, and has been given 3 months to implement an emergency action plan aimed at meeting the EU Air Quality Directive. The City of London has dangerous levels of PM10 concentrations, estimated to cause 3000 deaths per year within the city. As well as the threat of EU fines, in 2010 it was threatened with legal action for scrapping the western congestion charge zone, which is claimed to have led to an increase in air pollution levels.

In response to these charges, Boris Johnson, Mayor of London, has criticised the current need for European cities to communicate with Europe through their nation state's central government, arguing that in future "A great city like London" should be permitted to bypass its government and deal directly with the European Commission regarding its air quality action plan.

This can be interpreted as recognition that cities can transcend the traditional national government organisational hierarchy and develop solutions to air pollution using global governance networks, for example through transnational relations. Transnational relations include but are not exclusive to national governments and intergovernmental organisations, allowing sub-national actors including cities and regions to partake in air pollution control as independent actors.

Particularly promising at present are global city partnerships. These can be built into networks, for example the C40 Cities Climate Leadership Group, of which London is a member. The C40 is a public 'non-state' network of the world's leading cities that aims to curb their greenhouse emissions. The C40 has been identified as 'governance from the middle' and is an alternative to intergovernmental policy. It has the potential to improve urban air quality as participating cities "exchange information, learn from best practices and consequently mitigate carbon dioxide emissions independently from national government decisions". A criticism of the C40 network is that its exclusive nature limits influence to participating cities and risks drawing resources away from less powerful city and regional actors.

Atmospheric Dispersion

The basic technology for analyzing air pollution is through the use of a variety of mathematical models for predicting the transport of air pollutants in the lower atmosphere. The principal methodologies are:

- Point source dispersion, used for industrial sources

- Line source dispersion, used for airport and roadway air dispersion modeling

- Area source dispersion, used for forest fires or duststorms

- Photochemical models, used to analyze reactive pollutants that form smog

The point source problem is the best understood, since it involves simpler mathematics and has been studied for a long period of time, dating back to about the year 1900. It uses a Gaussian dispersion model for continuous buoyant pollution plumes to predict the air pollution isopleths, with consideration given to wind velocity, stack height, emission rate and stability class (a measure of atmospheric turbulence). This model has been extensively validated and calibrated with experimental data for all sorts of atmospheric conditions.

The roadway air dispersion model was developed starting in the late 1950s and early

1960s in response to requirements of the National Environmental Policy Act and the U.S. Department of Transportation (then known as the Federal Highway Administration) to understand impacts of proposed new highways upon air quality, especially in urban areas. Several research groups were active in this model development, among which were: the Environmental Research and Technology (ERT) group in Lexington, Massachusetts, the ESL Inc. group in Sunnyvale, California and the California Air Resources Board group in Sacramento, California. The research of the ESL group received a boost with a contract award from the United States Environmental Protection Agency to validate a line source model using sulfur hexafluoride as a tracer gas. This program was successful in validating the line source model developed by ESL Inc. Some of the earliest uses of the model were in court cases involving highway air pollution; the Arlington, Virginia portion of Interstate 66 and the New Jersey Turnpike widening project through East Brunswick, New Jersey.

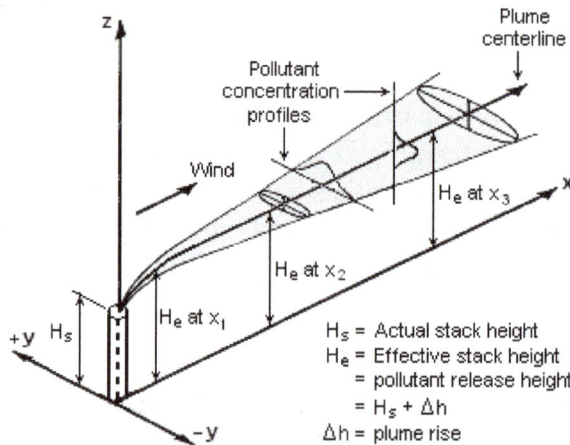

Visualization of a buoyant Gaussian air pollution dispersion plume as used in many atmospheric dispersion models.

Area source models were developed in 1971 through 1974 by the ERT and ESL groups, but addressed a smaller fraction of total air pollution emissions, so that their use and need was not as widespread as the line source model, which enjoyed hundreds of different applications as early as the 1970s. Similarly photochemical models were developed primarily in the 1960s and 70s, but their use was more specialized and for regional needs, such as understanding smog formation in Los Angeles, California.

Greenhouse Gas

A greenhouse gas (abbrev. GHG) is a gas in an atmosphere that absorbs and emits radiation within the thermal infrared range. This process is the fundamental cause of the greenhouse effect. The primary greenhouse gases in Earth's atmosphere are water vapor, carbon dioxide, methane, nitrous oxide, and ozone. Without greenhouse gases, the average temperature of Earth's surface would be about −18 °C (0 °F), rather than the present average of 15 °C (59 °F). In the Solar System, the atmospheres of Venus, Mars and Titan also contain gases that cause a greenhouse effect.

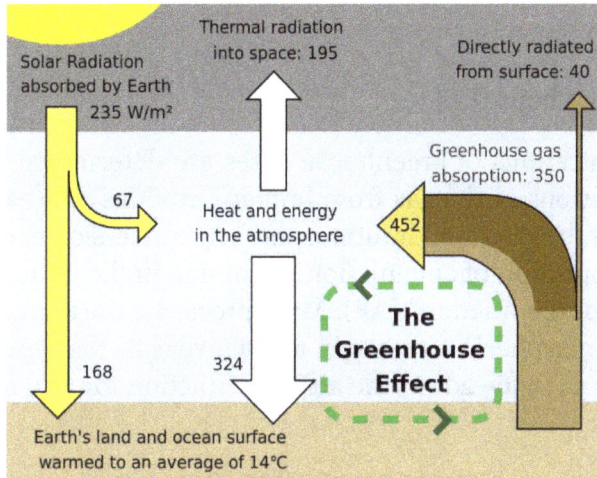

Greenhouse effect schematic showing energy flows between space, the atmosphere, and Earth's surface. Energy influx and emittance are expressed in watts per square meter (W/m²).

Human activities since the beginning of the Industrial Revolution (taken as the year 1750) have produced a 40% increase in the atmospheric concentration of carbon dioxide, from 280 ppm in 1750 to 400 ppm in 2015. This increase has occurred despite the uptake of a large portion of the emissions by various natural "sinks" involved in the carbon cycle. Anthropogenic carbon dioxide (CO_2) emissions (i.e. emissions produced by human activities) come from combustion of carbon-based fuels, principally coal, oil, and natural gas, along with deforestation, soil erosion and animal agriculture.

It has been estimated that if greenhouse gas emissions continue at the present rate, Earth's surface temperature could exceed historical values as early as 2047, with potentially harmful effects on ecosystems, biodiversity and the livelihoods of people worldwide. Recent estimates suggest that on the current emissions trajectory the Earth could pass a threshold of 2°C global warming, which the United Nations' IPCC designated as the upper limit for "dangerous" global warming, by 2036.

Gases in Earth's Atmosphere

Greenhouse Gases

Greenhouse gases are those that absorb and emit infrared radiation in the wavelength range emitted by Earth. In order, the most abundant greenhouse gases in Earth's atmosphere are:

- Water vapor (H2O)

- Carbon dioxide (CO_2)

- Methane (CH4)

- Nitrous oxide (N2O)

- Ozone (O3)

- Chlorofluorocarbons (CFCs)

Atmospheric concentrations of greenhouse gases are determined by the balance between sources (emissions of the gas from human activities and natural systems) and sinks (the removal of the gas from the atmosphere by conversion to a different chemical compound). The proportion of an emission remaining in the atmosphere after a specified time is the "airborne fraction" (AF). More precisely, the annual airborne fraction is the ratio of the atmospheric increase in a given year to that year's total emissions. Over the last 50 years (1956–2006) the airborne fraction for CO_2 has been increasing at $0.25 \pm 0.21\%$/year.

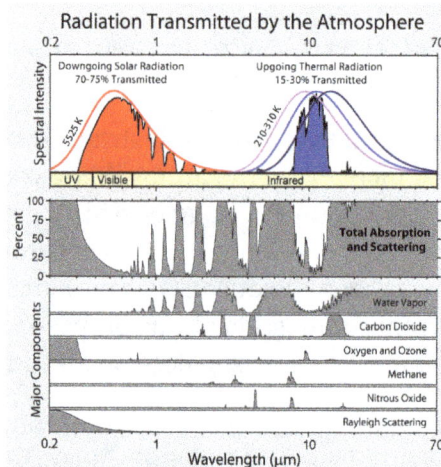

Atmospheric absorption and scattering at different wavelengths of electromagnetic waves. The largest absorption band of carbon dioxide is in the infrared.

Non-greenhouse Gases

The major atmospheric constituents, nitrogen (N2), oxygen (O2), and argon (Ar), are not greenhouse gases because molecules containing two atoms of the same element such as N2 and O2 and monatomic molecules such as argon (Ar) have no net change in the distribution of their electrical charges when they vibrate and hence are almost totally unaffected by infrared radiation. Although molecules containing two atoms of different elements such as carbon monoxide (CO) or hydrogen chloride (HCl) absorb infrared radiation, these molecules are short-lived in the atmosphere owing to their reactivity and solubility. Therefore, they do not contribute significantly to the greenhouse effect and usually are omitted when discussing greenhouse gases.

Indirect Radiative Effects

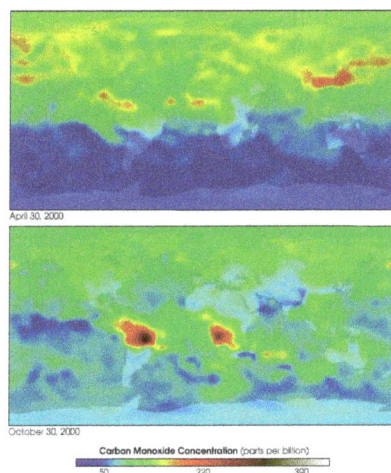

April 30, 2000

October 30, 2000

Carbon Monoxide Concentration (parts per billion)
50   220   390

The false colors in this image represent concentrations of carbon monoxide in the lower atmosphere, ranging from about 390 parts per billion (dark brown pixels), to 220 parts per billion (red pixels), to 50 parts per billion (blue pixels).

Some gases have indirect radiative effects (whether or not they are greenhouse gases themselves). This happens in two main ways. One way is that when they break down in the atmosphere they produce another greenhouse gas. For example, methane and carbon monoxide (CO) are oxidized to give carbon dioxide (and methane oxidation also produces water vapor; that will be considered below). Oxidation of CO to CO_2 directly produces an unambiguous increase in radiative forcing although the reason is subtle. The peak of the thermal IR emission from Earth's surface is very close to a strong vibrational absorption band of CO_2 (667 cm^{-1}). On the other hand, the single CO vibrational band only absorbs IR at much higher frequencies (2145 cm^{-1}), where the ~300 K thermal emission of the surface is at least a factor of ten lower. On the other hand, oxidation of methane to CO_2, which requires reactions with the OH radical, produces an instantaneous reduction, since CO_2 is a weaker greenhouse gas than methane; but it has a longer lifetime. As described below this is not the whole story, since the oxidations of CO and CH 4 are intertwined by both consuming OH radicals. In any case, the calculation of the total radiative effect needs to include both the direct and indirect forcing.

A second type of indirect effect happens when chemical reactions in the atmosphere involving these gases change the concentrations of greenhouse gases. For example, the destruction of non-methane volatile organic compounds (NMVOCs) in the atmosphere can produce ozone. The size of the indirect effect can depend strongly on where and when the gas is emitted.

Methane has a number of indirect effects in addition to forming CO_2. Firstly, the main chemical that destroys methane in the atmosphere is the hydroxyl radical (OH). Methane reacts with OH and so more methane means that the concentration of OH goes down. Effectively, methane increases its own atmospheric lifetime and therefore its overall radiative effect. The second effect is that the oxidation of methane can produce ozone. Thirdly, as well as making CO_2 the oxidation of methane produces water; this is a major source of water vapor in the stratosphere, which is otherwise very dry. CO and NMVOC also produce CO_2 when they are oxidized. They remove OH from the atmosphere and this leads to higher concentrations of methane. The surprising effect of this is that the global warming potential of CO is three times that of CO_2. The same process that converts NMVOC to carbon dioxide can also lead to the formation of tropospheric ozone. Halocarbons have an indirect effect because they destroy stratospheric ozone. Finally hydrogen can lead to ozone production and CH 4 increases as well as producing water vapor in the stratosphere.

Contribution of Clouds to Earth's Greenhouse Effect

The major non-gas contributor to Earth's greenhouse effect, clouds, also absorb and emit infrared radiation and thus have an effect on radiative properties of the greenhouse gases. Clouds are water droplets or ice crystals suspended in the atmosphere.

Impacts on the Overall Greenhouse Effect

Schmidt *et al.* (2010) analysed how individual components of the atmosphere contribute to the total greenhouse effect. They estimated that water vapor accounts for about 50% of Earth's greenhouse effect, with clouds contributing 25%, carbon dioxide 20%, and the minor greenhouse gases and aerosols accounting for the remaining 5%. In the study, the reference model atmosphere is for 1980 conditions. Image credit: NASA.

The contribution of each gas to the greenhouse effect is affected by the characteristics of that gas, its abundance, and any indirect effects it may cause. For example, the direct radiative effect of a mass of methane is about 72 times stronger than the same mass of carbon dioxide over a 20-year time frame but it is present in much smaller concentrations so that its total direct radiative effect is smaller, in part due to its shorter atmospheric lifetime. On the other hand, in addition to its direct radiative impact, methane has a large, indirect radiative effect because it contributes to ozone formation. Shindell *et al.* (2005) argue that the contribution to climate change from methane is at least double previous estimates as a result of this effect.

When ranked by their direct contribution to the greenhouse effect, the most important are:

Compound	Formula	Concentration in atmosphere (ppm)	Contribution (%)
Water vapor and clouds	H_2O	10–50,000[A]	36–72%
Carbon dioxide	CO_2	~400	9–26%
Methane	CH_4	~1.8	4–9%

Ozone	O 3	2–8[B]	3–7%
notes: [A] Water vapor strongly varies locally [B] The concentration in stratosphere. About 90% of the ozone in Earth's atmosphere is contained in the stratosphere.			

In addition to the main greenhouse gases listed above, other greenhouse gases include sulfur hexafluoride, hydrofluorocarbons and perfluorocarbons. Some greenhouse gases are not often listed. For example, nitrogen trifluoride has a high global warming potential (GWP) but is only present in very small quantities.

Proportion of Direct Effects at a Given Moment

It is not possible to state that a certain gas causes an exact percentage of the greenhouse effect. This is because some of the gases absorb and emit radiation at the same frequencies as others, so that the total greenhouse effect is not simply the sum of the influence of each gas. The higher ends of the ranges quoted are for each gas alone; the lower ends account for overlaps with the other gases. In addition, some gases such as methane are known to have large indirect effects that are still being quantified.

Atmospheric Lifetime

Aside from water vapor, which has a residence time of about nine days, major greenhouse gases are well mixed and take many years to leave the atmosphere. Although it is not easy to know with precision how long it takes greenhouse gases to leave the atmosphere, there are estimates for the principal greenhouse gases. Jacob (1999) defines the lifetime τ of an atmospheric species X in a one-box model as the average time that a molecule of X remains in the box. Mathematically τ can be defined as the ratio of the mass m (in kg) of X in the box to its removal rate, which is the sum of the flow of X out of the box (F_{out}), chemical loss of X (D), and deposition of X (D) (all in kg/s):

$\tau = \dfrac{m}{F_{out} + L + D}$. If one stopped pouring any of this gas into the box, then after a time τ, its concentration would be about halved.

The atmospheric lifetime of a species therefore measures the time required to restore equilibrium following a sudden increase or decrease in its concentration in the atmosphere. Individual atoms or molecules may be lost or deposited to sinks such as the soil, the oceans and other waters, or vegetation and other biological systems, reducing the excess to background concentrations. The average time taken to achieve this is the mean lifetime.

Carbon dioxide has a variable atmospheric lifetime, and cannot be specified precisely. The atmospheric lifetime of CO_2 is estimated of the order of 30–95 years. This figure accounts for CO_2 molecules being removed from the atmosphere by mixing into the ocean, photosynthesis, and other processes. However, this excludes the balancing fluxes of CO_2 into the atmosphere from the geological reservoirs, which have slower characteristic rates. Although more than half of the CO_2 emitted is removed from the atmosphere within a century, some fraction (about 20%) of emitted CO_2 remains in the atmosphere for many thousands of years. Similar issues apply to other greenhouse gases, many of which have longer mean lifetimes than CO_2. E.g., N_2O has a mean atmospheric lifetime of 114 years.

Radiative Forcing

Earth absorbs some of the radiant energy received from the sun, reflects some of it as light and reflects or radiates the rest back to space as heat. Earth's surface temperature depends on this balance between incoming and outgoing energy. If this energy balance is shifted, Earth's surface could become warmer or cooler, leading to a variety of changes in global climate.

A number of natural and man-made mechanisms can affect the global energy balance and force changes in Earth's climate. Greenhouse gases are one such mechanism. Greenhouse gases in the atmosphere absorb and re-emit some of the outgoing energy radiated from Earth's surface, causing that heat to be retained in the lower atmosphere. As explained above, some greenhouse gases remain in the atmosphere for decades or even centuries, and therefore can affect Earth's energy balance over a long time period. Factors that influence Earth's energy balance can be quantified in terms of "radiative climate forcing." Positive radiative forcing indicates warming (for example, by increasing incoming energy or decreasing the amount of energy that escapes to space), whereas negative forcing is associated with cooling.

Global Warming Potential

The global warming potential (GWP) depends on both the efficiency of the molecule as a greenhouse gas and its atmospheric lifetime. GWP is measured relative to the same mass of CO_2 and evaluated for a specific timescale. Thus, if a gas has a high (positive) radiative forcing but also a short lifetime, it will have a large GWP on a 20-year scale but a small one on a 100-year scale. Conversely, if a molecule has a longer atmospheric lifetime than CO_2 its GWP will increase with the timescale considered. Carbon dioxide is defined to have a GWP of 1 over all time periods.

Methane has an atmospheric lifetime of 12 ± 3 years. The 2007 IPCC report lists the GWP as 72 over a time scale of 20 years, 25 over 100 years and 7.6 over 500 years. A 2014 analysis, however, states that although methane's initial impact is about 100 times greater than that of CO_2, because of the shorter atmospheric lifetime, after six or seven decades, the impact of the two gases is about equal, and from then on methane's relative role continues to decline. The decrease in GWP at longer times is because methane is degraded to water and CO_2 through chemical reactions in the atmosphere.

Examples of the atmospheric lifetime and GWP relative to CO_2 for several greenhouse gases are given in the following table:

Atmospheric lifetime and GWP relative to CO_2 at different time horizon for various greenhouse gases.					
Gas name	**Chemical formula**	**Lifetime (years)**	**Global warming potential (GWP) for given time horizon**		
			20-yr	**100-yr**	**500-yr**
Carbon dioxide	CO_2	30–95	1	1	1
Methane	CH4	12	72	25	7.6
Nitrous oxide	N2O	114	289	298	153
CFC-12	CCl2F2	100	11 000	10 900	5 200
HCFC-22	CHClF2	12	5 160	1 810	549
Tetrafluoromethane	CF4	50 000	5 210	7 390	11 200
Hexafluoroethane	C2F6	10 000	8 630	12 200	18 200
Sulfur hexafluoride	SF6	3 200	16 300	22 800	32 600
Nitrogen trifluoride	NF3	740	12 300	17 200	20 700

The use of CFC-12 (except some essential uses) has been phased out due to its ozone depleting properties. The phasing-out of less active HCFC-compounds will be completed in 2030.

Carbon dioxide in Earth's atmosphere if *half* of global-warming emissions are *not* absorbed.
(NASA simulation; 9 November 2015)

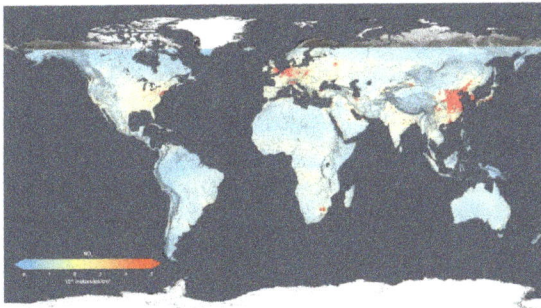

Nitrogen dioxide 2014 – global air quality levels
(released 14 December 2015).

Natural and Anthropogenic Sources

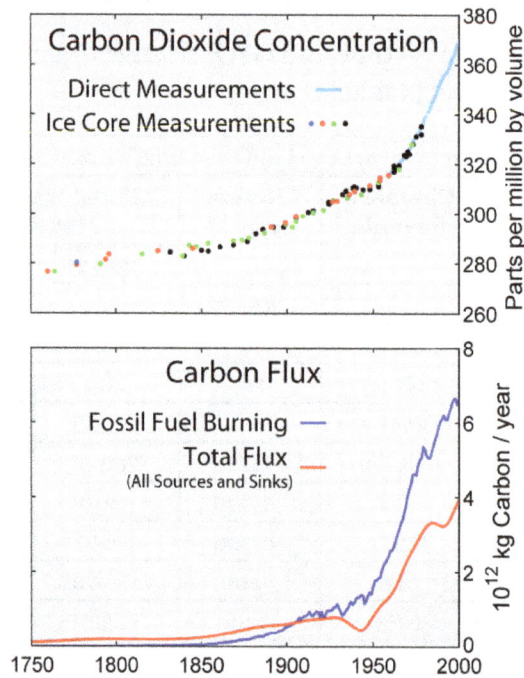

Top: Increasing atmospheric carbon dioxide levels as measured in the atmosphere and reflected in ice cores. Bottom: The amount of net carbon increase in the atmosphere, compared to carbon emissions from burning fossil fuel.

This diagram shows a simplified representation of the contemporary global carbon cycle. Changes are measured in gigatons of carbon per year (GtC/y). Canadell *et al.* (2007) estimated the growth rate of global average atmospheric CO_2 for 2000–2006 as 1.93 parts-per-million per year (4.1 petagrams of carbon per year).

Aside from purely human-produced synthetic halocarbons, most greenhouse gases have both natural and human-caused sources. During the pre-industrial Holocene, concentrations of existing gases were roughly constant. In the industrial era, human activities have added greenhouse gases to the atmosphere, mainly through the burning of fossil fuels and clearing of forests.

The 2007 Fourth Assessment Report compiled by the IPCC (AR4) noted that "changes in atmospheric concentrations of greenhouse gases and aerosols, land cover and solar radiation alter the energy balance of the climate system", and concluded that "increases in anthropogenic greenhouse gas concentrations is very likely to have caused most of the increases in global average temperatures since the mid-20th century". In AR4, "most of" is defined as more than 50%.

Abbreviations used in the two tables below: ppm = parts-per-million; ppb = parts-per-billion; ppt = parts-per-trillion; W/m² = watts per square metre

Current greenhouse gas concentrations					
Gas	Pre-1750 tropospheric concentration	Recent tropospheric concentration	Absolute increase since 1750	Percentage increase since 1750	Increased radiative forcing (W/m²)
Carbon dioxide (CO_2)	280 ppm	395.4 ppm	115.4 ppm	41.2%	1.88
Methane (CH4)	700 ppb	1893 ppb / 1762 ppb	1193 ppb / 1062 ppb	170.4% / 151.7%	0.49
Nitrous oxide (N2O)	270 ppb	326 ppb / 324 ppb	56 ppb / 54 ppb	20.7% / 20.0%	0.17
Tropospheric ozone (O3)	237 ppb	337 ppb	100 ppb	42%	0.4

Relevant to radiative forcing and/or ozone depletion; all of the following have no natural sources and hence zero amounts pre-industrial		
Gas	Recent tropospheric concentration	Increased radiative forcing (W/m²)
CFC-11 (trichlorofluoromethane) (CCl3F)	236 ppt / 234 ppt	0.061
CFC-12 (CCl2F2)	527 ppt / 527 ppt	0.169
CFC-113 (Cl2FC-CClF2)	74 ppt / 74 ppt	0.022
HCFC-22 (CHClF2)	231 ppt / 210 ppt	0.046
HCFC-141b (CH3CCl2F)	24 ppt / 21 ppt	0.0036
HCFC-142b (CH3CClF2)	23 ppt / 21 ppt	0.0042
Halon 1211 (CBrClF2)	4.1 ppt / 4.0 ppt	0.0012
Halon 1301 (CBrClF3)	3.3 ppt / 3.3 ppt	0.001
HFC-134a (CH2FCF3)	75 ppt / 64 ppt	0.0108
Carbon tetrachloride (CCl4)	85 ppt / 83 ppt	0.0143
Sulfur hexafluoride (SF6)	7.79 ppt / 7.39 ppt	0.0043
Other halocarbons	Varies by substance	collectively 0.02
Halocarbons in total		0.3574

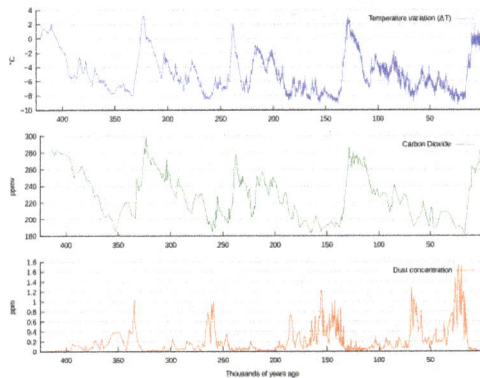

400,000 years of ice core data

Ice cores provide evidence for greenhouse gas concentration variations over the past 800,000 years. Both CO_2 and CH4 vary between glacial and interglacial phases, and concentrations of these gases correlate strongly with temperature. Direct data does not exist for periods earlier than those represented in the ice core record, a record that indicates CO_2 mole fractions stayed within a range of 180 ppm to 280 ppm throughout the last 800,000 years, until the increase of the last 250 years. However, various proxies and modeling suggests larger variations in past epochs; 500 million years ago CO_2 levels were likely 10 times higher than now. Indeed, higher CO_2 concentrations are thought to have prevailed throughout most of the Phanerozoic eon, with concentrations four to six times current concentrations during the Mesozoic era, and ten to fifteen times current concentrations during the early Palaeozoic era until the middle of the Devonian period, about 400 Ma. The spread of land plants is thought to have reduced CO_2 concentrations during the late Devonian, and plant activities as both sources and sinks of CO_2 have since been important in providing stabilising feedbacks. Earlier still, a 200-million year period of intermittent, widespread glaciation extending close to the equator (Snowball Earth) appears to have been ended suddenly, about 550 Ma, by a colossal volcanic outgassing that raised the CO_2 concentration of the atmo-sphere abruptly to 12%, about 350 times modern levels, causing extreme greenhouse conditions and carbonate deposition as limestone at the rate of about 1 mm per day. This episode marked the close of the Precambrian eon, and was succeeded by the gen-erally warmer conditions of the Phanerozoic, during which multicellular animal and plant life evolved. No volcanic carbon dioxide emission of comparable scale has oc-curred since. In the modern era, emissions to the atmosphere from volcanoes are only about 1% of emissions from human sources.

Ice Cores

Measurements from Antarctic ice cores show that before industrial emissions started atmospheric CO_2 mole fractions were about 280 parts per million (ppm), and stayed between 260 and 280 during the preceding ten thousand years. Carbon dioxide mole fractions in the atmosphere have gone up by approximately 35 percent since the 1900s, rising from 280 parts per million by volume to 387 parts per million in 2009. One study using evidence from stomata of fossilized leaves suggests greater variability, with carbon dioxide mole fractions above 300 ppm during the period seven to ten thousand years ago, though others have argued that these findings more likely reflect calibration or contam-ination problems rather than actual CO_2 variability. Because of the way air is trapped in ice (pores in the ice close off slowly to form bubbles deep within the firn) and the time period represented in each ice sample analyzed, these figures represent averages of at-mospheric concentrations of up to a few centuries rather than annual or decadal levels.

Changes Since the Industrial Revolution

Since the beginning of the Industrial Revolution, the concentrations of most of the greenhouse gases have increased. For example, the mole fraction of carbon dioxide has increased from 280 ppm by about 36% to 380 ppm, or 100 ppm over modern pre-industrial levels. The first 50 ppm increase took place in about 200 years, from the start of the Industrial Revolution to around 1973.; however the next 50 ppm increase took place in about 33 years, from 1973 to 2006.

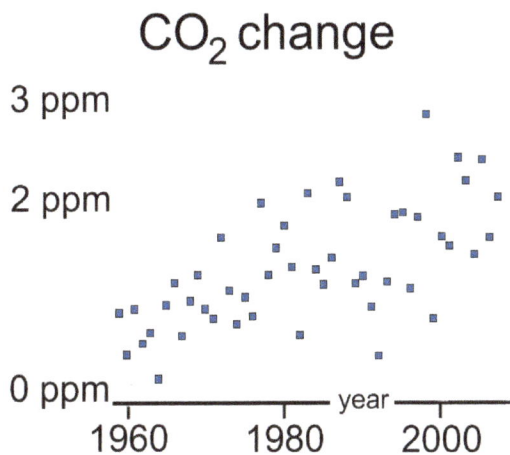

Recent year-to-year increase of atmospheric CO_2.

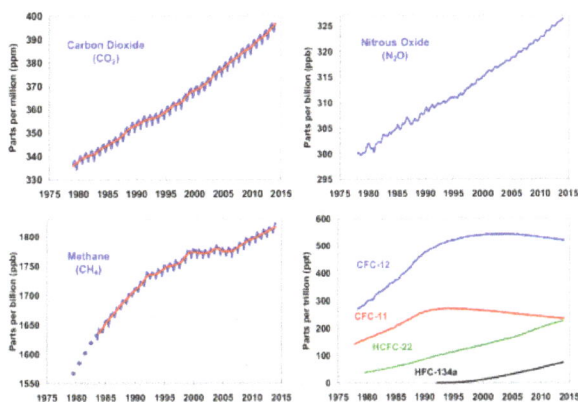

Major greenhouse gas trends.

ecent data also shows that the concentration is increasing at a higher rate. In the 1960s, the average annual increase was only 37% of what it was in 2000 through 2007.

Today, the stock of carbon in the atmosphere increases by more than 3 million tonnes per annum (0.04%) compared with the existing stock. This increase is the result of human activities by burning fossil fuels, deforestation and forest degradation in tropical and boreal regions.

The other greenhouse gases produced from human activity show similar increases in both amount and rate of increase. Many observations are available online in a variety of Atmospheric Chemistry Observational Databases.

Anthropogenic Greenhouse Gases

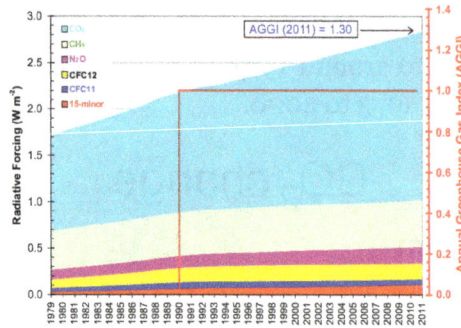

This graph shows changes in the annual greenhouse gas index (AGGI) between 1979 and 2011. The AGGI measures the levels of greenhouse gases in the atmosphere based on their ability to cause changes in Earth's climate.

This bar graph shows global greenhouse gas emissions by sector from 1990 to 2005, measured in carbon dioxide equivalents.

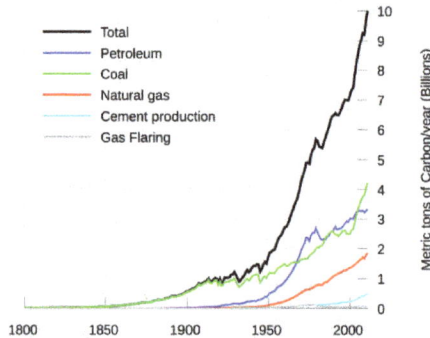

Modern global CO2 emissions from the burning of fossil fuels.

Since about 1750 human activity has increased the concentration of carbon dioxide and other greenhouse gases. Measured atmospheric concentrations of carbon dioxide are currently 100 ppm higher than pre-industrial levels. Natural sources of carbon dioxide are more than 20 times greater than sources due to human activity, but over periods longer than a few years natural sources are closely balanced by natural sinks, mainly photosynthesis of carbon compounds by plants and marine plankton. As a result of this balance, the atmospheric mole fraction of carbon dioxide remained between 260 and 280 parts per million for the 10,000 years between the end of the last glacial maximum and the start of the industrial era.

It is likely that anthropogenic (i.e., human-induced) warming, such as that due to elevated greenhouse gas levels, has had a discernible influence on many physical and biological systems. Future warming is projected to have a range of impacts, including sea level rise, increased frequencies and severities of some extreme weather events, loss of biodiversity, and regional changes in agricultural productivity.

The main sources of greenhouse gases due to human activity are:

- burning of fossil fuels and deforestation leading to higher carbon dioxide concentrations in the air. Land use change (mainly deforestation in the tropics) account for up to one third of total anthropogenic CO_2 emissions.

- livestock enteric fermentation and manure management, paddy rice farming, land use and wetland changes, pipeline losses, and covered vented landfill emissions leading to higher methane atmospheric concentrations. Many of the newer style fully vented septic systems that enhance and target the fermentation process also are sources of atmospheric methane.

- use of chlorofluorocarbons (CFCs) in refrigeration systems, and use of CFCs and halons in fire suppression systems and manufacturing processes.

- agricultural activities, including the use of fertilizers, that lead to higher nitrous oxide (N2O) concentrations.

The seven sources of CO_2 from fossil fuel combustion are (with percentage contributions for 2000–2004):

Seven main fossil fuel combustion sources	Contribution (%)
Liquid fuels (e.g., gasoline, fuel oil)	36%
Solid fuels (e.g., coal)	35%
Gaseous fuels (e.g., natural gas)	20%
Cement production	3 %
Flaring gas industrially and at wells	< 1%
Non-fuel hydrocarbons	< 1%
"International bunker fuels" of transport not included in national inventories	4 %

Carbon dioxide, methane, nitrous oxide (N2O) and three groups of fluorinated gases (sulfur hexafluoride (SF6), hydrofluorocarbons (HFCs), and perfluorocarbons (PFCs)) are the major anthropogenic greenhouse gases, and are regulated under the Kyoto Protocol international treaty, which came into force in 2005. Emissions limitations specified in the Kyoto Protocol expired in 2012. The Cancún agreement, agreed in 2010, includes voluntary pledges made by 76 countries to control emissions. At the time of the agreement, these 76 countries were collectively responsible for 85% of annual global emissions.

Although CFCs are greenhouse gases, they are regulated by the Montreal Protocol, which was motivated by CFCs' contribution to ozone depletion rather than by their contribution to global warming. Note that ozone depletion has only a minor role in greenhouse warming though the two processes often are confused in the media. On 15 October 2016, negotiators from over 170 nations meeting at the summit of the United

Nations Environment Programme reached a legally-binding accord to phase out hydro-fluorocarbons (HFCs) in an amendment to the Montreal Protocol.

Sectors

Tourism

According to UNEP global tourism is closely linked to climate change. Tourism is a significant contributor to the increasing concentrations of greenhouse gases in the atmosphere. Tourism accounts for about 50% of traffic movements. Rapidly expanding air traffic contributes about 2.5% of the production of CO_2. The number of international travelers is expected to increase from 594 million in 1996 to 1.6 billion by 2020, adding greatly to the problem unless steps are taken to reduce emissions.

Road Haulage

The road haulage industry plays a part in production of CO_2, contributing around 20% of the UK's total carbon emissions a year, with only the energy industry having a larger impact at around 39%. Average carbon emissions within the haulage industry are falling—in the thirty-year period from 1977–2007, the carbon emissions associated with a 200-mile journey fell by 21 percent; NOx emissions are also down 87 percent, whereas journey times have fallen by around a third. Due to their size, HGVs often receive criticism regarding their CO2 emissions; however, rapid development in engine technology and fuel management is having a largely positive effect.

Role of Water Vapor

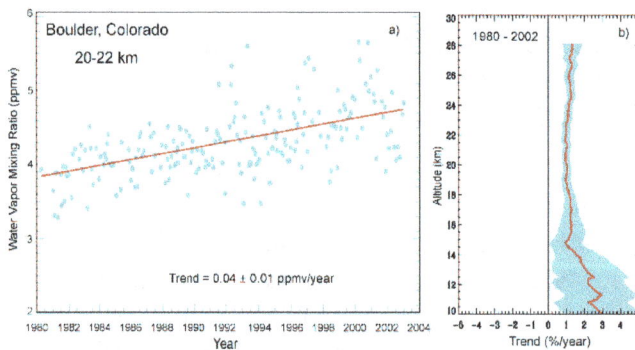

Increasing water vapor in the stratosphere at Boulder, Colorado.

Water vapor accounts for the largest percentage of the greenhouse effect, between 36% and 66% for clear sky conditions and between 66% and 85% when including clouds. Water vapor concentrations fluctuate regionally, but human activity does not significantly affect water vapor concentrations except at local scales, such as near irrigated fields. The atmospheric concentration of vapor is highly variable and depends largely on temperature, from less than 0.01% in extremely cold regions up to 3% by mass in saturated air at about 32 °C.

The average residence time of a water molecule in the atmosphere is only about nine days, compared to years or centuries for other greenhouse gases such as CH 4 and CO_2. Thus, water vapor responds to and amplifies effects of the other greenhouse gases. The Clausius–Clapeyron relation establishes that more water vapor will be present per unit volume at elevated temperatures. This and other basic principles indicate that warming associated with increased concentrations of the other greenhouse gases also will increase the concentration of water vapor (assuming that the relative humidity remains approximately constant; modeling and observational studies find that this is indeed so). Because water vapor is a greenhouse gas, this results in further warming and so is a "positive feedback" that amplifies the original warming. Eventually other earth processes offset these positive feedbacks, stabilizing the global temperature at a new equilibrium and preventing the loss of Earth's water through a Venus-like runaway greenhouse effect.

Direct Greenhouse Gas Emissions

Between the period 1970 to 2004, GHG emissions (measured in CO_2-equivalent) increased at an average rate of 1.6% per year, with CO_2 emissions from the use of fossil fuels growing at a rate of 1.9% per year. Total anthropogenic emissions at the end of 2009 were estimated at 49.5 gigatonnes CO_2-equivalent. These emissions include CO_2 from fossil fuel use and from land use, as well as emissions of methane, nitrous oxide and other GHGs covered by the Kyoto Protocol.

At present, the primary source of CO_2 emissions is the burning of coal, natural gas, and petroleum for electricity and heat.

Regional and National Attribution of Emissions

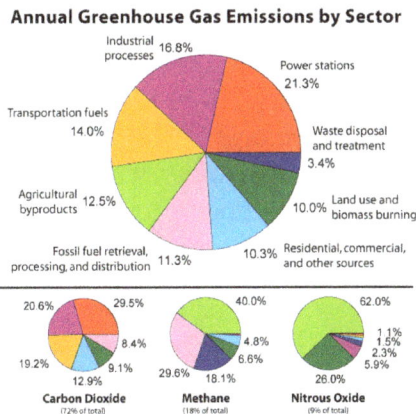

Annual Greenhouse Gas Emissions by Sector

Industrial processes 16.8%
Power stations 21.3%
Transportation fuels 14.0%
Waste disposal and treatment 3.4%
Agricultural byproducts 12.5%
Land use and biomass burning 10.0%
Fossil fuel retrieval, processing, and distribution 11.3%
Residential, commercial, and other sources 10.3%

Carbon Dioxide (72% of total): 29.5%, 8.4%, 9.1%, 12.9%, 19.2%, 20.6%

Methane (18% of total): 40.0%, 4.8%, 6.6%, 18.1%, 29.6%

Nitrous Oxide (9% of total): 62.0%, 1.1%, 1.5%, 2.3%, 5.9%, 26.0%

This figure shows the relative fraction of anthropogenic greenhouse gases coming from each of eight categories of sources, as estimated by the Emission Database for Global

Atmospheric Research version 3.2, fast track 2000 project . These values are intended to provide a snapshot of global annual greenhouse gas emissions in the year 2000. The top panel shows the sum over all anthropogenic greenhouse gases, weighted by their global warming potential over the next 100 years. This consists of 72% carbon dioxide, 18% methane, 8% nitrous oxide and 1% other gases. Lower panels show the comparable information for each of these three primary greenhouse gases, with the same coloring of sectors as used in the top chart. Segments with less than 1% fraction are not labeled.

There are several different ways of measuring GHG emissions, for example, World Bank (2010) for tables of national emissions data. Some variables that have been reported include:

- Definition of measurement boundaries: Emissions can be attributed geographically, to the area where they were emitted (the territory principle) or by the activity principle to the territory produced the emissions. These two principles result in different totals when measuring, for example, electricity importation from one country to another, or emissions at an international airport.

- Time horizon of different GHGs: Contribution of a given GHG is reported as a CO_2 equivalent. The calculation to determine this takes into account how long that gas remains in the atmosphere. This is not always known accurately and calculations must be regularly updated to reflect new information.

- What sectors are included in the calculation (e.g., energy industries, industrial processes, agriculture etc.): There is often a conflict between transparency and availability of data.

- The measurement protocol itself: This may be via direct measurement or estimation. The four main methods are the emission factor-based method, mass balance method, predictive emissions monitoring systems, and continuous emissions monitoring systems. These methods differ in accuracy, cost, and usability.

These different measures are sometimes used by different countries to assert various policy/ethical positions on climate change (Banuri *et al.*, 1996, p. 94). This use of different measures leads to a lack of comparability, which is problematic when monitoring progress towards targets. There are arguments for the adoption of a common measurement tool, or at least the development of communication between different tools.

Emissions may be measured over long time periods. This measurement type is called historical or cumulative emissions. Cumulative emissions give some indication of who is responsible for the build-up in the atmospheric concentration of GHGs (IEA, 2007, p. 199).

The national accounts balance would be positively related to carbon emissions. The national accounts balance shows the difference between exports and imports. For many richer nations, such as the United States, the accounts balance is negative because more goods are imported than they are exported. This is mostly due to the fact that it is cheaper to produce goods outside of developed countries, leading the economies of developed countries to become increasingly dependent on services and not goods. We believed that a positive accounts balance would means that more production was occurring in a country, so more factories working would increase carbon emission levels. (Holtz-Eakin, 1995, pp.;85;101).

Emissions may also be measured across shorter time periods. Emissions changes may, for example, be measured against a base year of 1990. 1990 was used in the United Nations Framework Convention on Climate Change (UNFCCC) as the base year for emissions, and is also used in the Kyoto Protocol (some gases are also measured from the year 1995). A country's emissions may also be reported as a proportion of global emissions for a particular year.

Another measurement is of per capita emissions. This divides a country's total annual emissions by its mid-year population. Per capita emissions may be based on historical or annual emissions (Banuri *et al.*, 1996, pp. 106–107).

Land-use Change

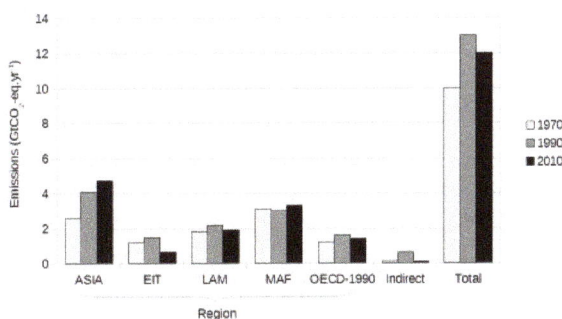

Greenhouse gas emissions from agriculture, forestry and other land use, 1970–2010.

Land-use change, e.g., the clearing of forests for agricultural use, can affect the concentration of GHGs in the atmosphere by altering how much carbon flows out of the atmosphere into carbon sinks. Accounting for land-use change can be understood as an attempt to measure "net" emissions, i.e., gross emissions from all GHG sources minus the removal of emissions from the atmosphere by carbon sinks (Banuri *et al.*, 1996, pp. 92–93).

There are substantial uncertainties in the measurement of net carbon emissions. Additionally, there is controversy over how carbon sinks should be allocated between different regions and over time (Banuri *et al.*, 1996, p. 93). For instance, concentrating on more recent changes in carbon sinks is likely to favour those regions that have deforested earlier, e.g., Europe.

Greenhouse Gas Intensity

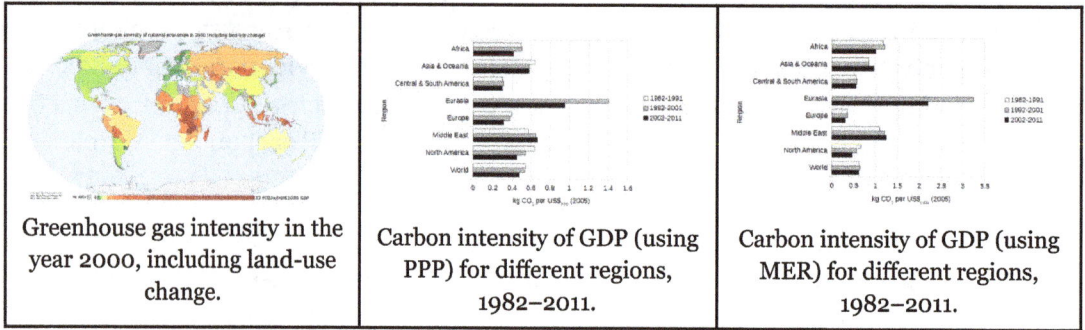

Greenhouse gas intensity in the year 2000, including land-use change.	Carbon intensity of GDP (using PPP) for different regions, 1982–2011.	Carbon intensity of GDP (using MER) for different regions, 1982–2011.

Greenhouse gas intensity is a ratio between greenhouse gas emissions and another metric, e.g., gross domestic product (GDP) or energy use. The terms "carbon intensity" and "emissions intensity" are also sometimes used. GHG intensities may be calculated using market exchange rates (MER) or purchasing power parity (PPP) (Banuri *et al.*, 1996, p. 96). Calculations based on MER show large differences in intensities between developed and developing countries, whereas calculations based on PPP show smaller differences.

Cumulative and Historical Emissions

Cumulative energy-related CO_2 emissions between the years 1850–2005 grouped into low-income, middle-income, high-income, the EU-15, and the OECD countries.

Cumulative energy-related CO_2 emissions between the years 1850–2005 for individual countries.

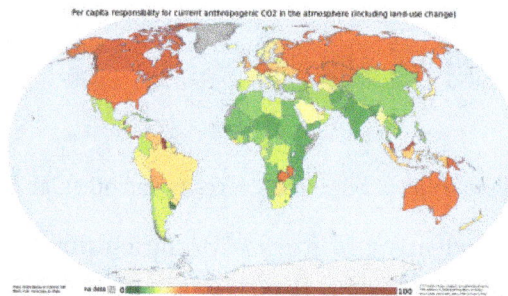

Map of cumulative per capita anthropogenic atmospheric CO_2 emissions by country. Cumulative emissions include land use change, and are measured between the years 1950 and 2000.

Regional trends in annual CO_2 emissions from fuel combustion between 1971 and 2009.

Regional trends in annual per capita CO_2 emissions from fuel combustion between 1971 and 2009.

Cumulative anthropogenic (i.e., human-emitted) emissions of CO_2 from fossil fuel use are a major cause of global warming, and give some indication of which countries have contributed most to human-induced climate change.

Top-5 historic CO_2 contributors by region over the years 1800 to 1988 (in %)		
Region	**Industrial CO_2**	**Total CO_2**
OECD North America	33.2	29.7
OECD Europe	26.1	16.6
Former USSR	14.1	12.5
China	5.5	6.0
Eastern Europe	5.5	4.8

The table above to the left is based on Banuri *et al.* (1996, p. 94). Overall, developed countries accounted for 83.8% of industrial CO_2 emissions over this time period, and 67.8% of total CO_2 emissions. Developing countries accounted for industrial CO_2 emissions of 16.2% over this time period, and 32.2% of total CO_2 emissions. The estimate of total CO_2 emissions includes biotic carbon emissions, mainly from deforestation. Banuri *et al.* (1996, p. 94) calculated per capita cumulative emissions based on then-current population. The ratio in per capita emissions between industrialized countries and developing countries was estimated at more than 10 to 1.

Including biotic emissions brings about the same controversy mentioned earlier regarding carbon sinks and land-use change (Banuri *et al.*, 1996, pp. 93–94). The actual calculation of net emissions is very complex, and is affected by how carbon sinks are allocated between regions and the dynamics of the climate system.

Non-OECD countries accounted for 42% of cumulative energy-related CO_2 emissions between 1890–2007. Over this time period, the US accounted for 28% of emis-sions; the EU, 23%; Russia, 11%; China, 9%; other OECD countries, 5%; Japan, 4%; India, 3%; and the rest of the world, 18%.

Changes Since a Particular Base Year

Between 1970–2004, global growth in annual CO_2 emissions was driven by North America, Asia, and the Middle East. The sharp acceleration in CO_2 emissions since 2000 to more than a 3% increase per year (more than 2 ppm per year) from 1.1% per year during the 1990s is attributable to the lapse of formerly declining trends in carbon intensity of both developing and developed nations. China was responsible for most of global growth in emissions during this period. Localised plummeting emissions associated with the collapse of the Soviet Union have been followed by slow emissions growth in this region due to more efficient energy use, made necessary by the increasing proportion of it that is exported. In comparison, methane has not increased appreciably, and N 2O by 0.25% y^{-1}.

Using different base years for measuring emissions has an effect on estimates of national contributions to global warming. This can be calculated by dividing a country's highest contribution to global warming starting from a particular base year, by that country's minimum contribution to global warming starting from a particular base year. Choosing between different base years of 1750, 1900, 1950, and 1990 has a significant effect for most countries.[17-18] Within the G8 group of countries, it is most significant for the UK, France and Germany. These countries have a long history of CO_2 emissions.

Annual Emissions

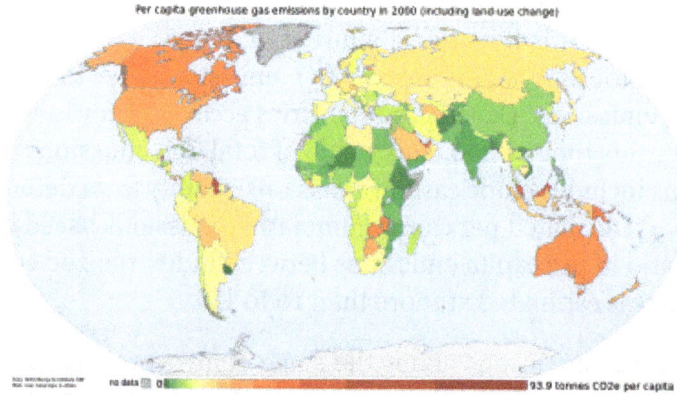

Per capita anthropogenic greenhouse gas emissions by country
for the year 2000 including land-use change.

Annual per capita emissions in the industrialized countries are typically as much as ten times the average in developing countries. Due to China's fast economic development, its annual per capita emissions are quickly approaching the levels of those in the Annex I group of the Kyoto Protocol (i.e., the developed countries excluding the USA). Other countries with fast growing emissions are South Korea, Iran, and Australia (which apart from the oil rich Persian Gulf states, now has the highest per-capita emission rate in the world). On the other hand, annual per capita emissions of the EU-15 and the USA are gradually decreasing over time. Emissions in Russia and Ukraine have decreased fastest since 1990 due to economic restructuring in these countries.

Energy statistics for fast growing economies are less accurate than those for the industrialized countries. For China's annual emissions in 2008, the Netherlands Environmental Assessment Agency estimated an uncertainty range of about 10%.

The GHG footprint, or greenhouse gas footprint, refers to the amount of GHG that are emitted during the creation of products or services. It is more comprehensive than the commonly used carbon footprint, which measures only carbon dioxide, one of many greenhouse gases.

2015 was the first year to see both total global economic growth and a reduction of carbon emissions.

Top Emitter Countries

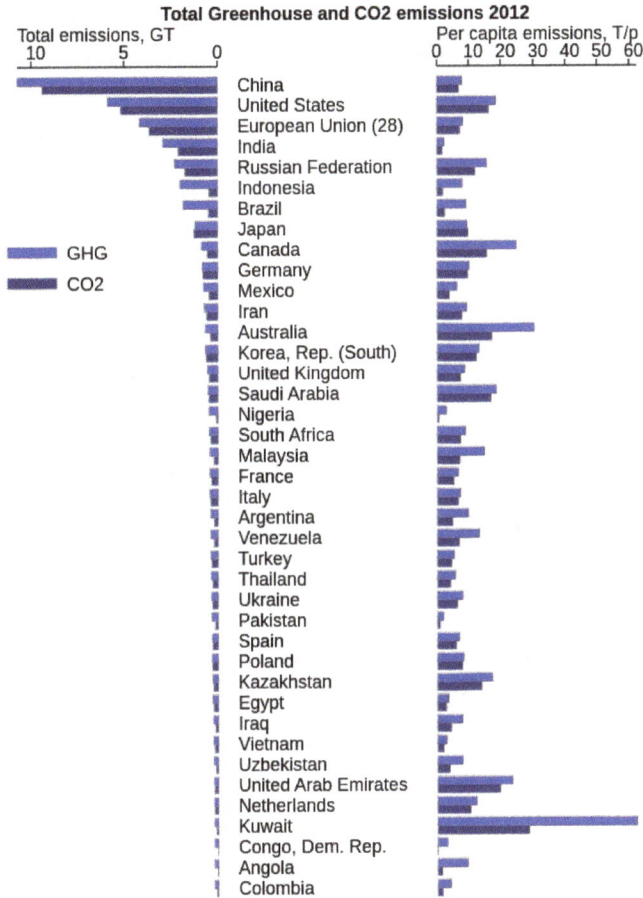

Total Greenhouse and CO2 emissions 2012

The top 40 countries emitting all greenhouse gases, showing both that derived from all sources including land clearance and forestry and also the CO2 component excluding those sources. Per capita figures are included. Data taken from World Resources Institute, Washington. Note that Indonesia and Brazil show very much higher than on graphs simply showing fossil fuel use.

Annual

In 2009, the annual top ten emitting countries accounted for about two-thirds of the world's annual energy-related CO_2 emissions.

Top-10 annual energy-related CO_2 emitters for the year 2009		
Country	**% of global total annual emissions**	**Tonnes of GHG per capita**
People's Rep. of China	23.6	5.1
United States	17.9	16.9

India	5.5	1.4
Russian Federation	5.3	10.8
Japan	3.8	8.6
Germany	2.6	9.2
Islamic Rep. of Iran	1.8	7.3
Canada	1.8	15.4
South Korea	1.8	10.6
United Kingdom	1.6	7.5

Cumulative

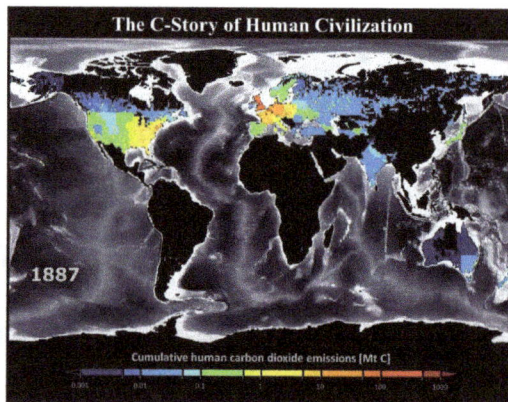

The C-Story of Human Civilization by PIK

Top-10 cumulative energy-related CO_2 emitters between 1850–2008		
Country	**% of world total**	**Metric tonnes CO_2 per person**
United States	28.5	1,132.7
China	9.36	85.4
Russian Federation	7.95	677.2
Germany	6.78	998.9
United Kingdom	5.73	1,127.8
Japan	3.88	367
France	2.73	514.9
India	2.52	26.7
Canada	2.17	789.2
Ukraine	2.13	556.4

Embedded Emissions

One way of attributing greenhouse gas (GHG) emissions is to measure the embedded emissions (also referred to as "embodied emissions") of goods that are being consumed. Emissions are usually measured according to production, rather than consumption. For example, in the main international treaty on climate change (the UNFCCC), countries report on emissions produced within their borders, e.g., the emissions produced from burning fossil fuels. Under a production-based accounting of emissions, embedded emissions on im-ported goods are attributed to the exporting, rather than the importing, country. Under a consumption-based accounting of emissions, embedded emissions on imported goods are attributed to the importing country, rather than the exporting, country.

Davis and Caldeira (2010) found that a substantial proportion of CO_2 emissions are traded internationally. The net effect of trade was to export emissions from China and other emerging markets to consumers in the US, Japan, and Western Europe. Based on annual emissions data from the year 2004, and on a per-capita consumption basis, the top-5 emitting countries were found to be (in tCO_2 per person, per year): Luxembourg (34.7), the US (22.0), Singapore (20.2), Australia (16.7), and Canada (16.6). Carbon Trust research revealed that approximately 25% of all CO_2 emissions from hu-man activities 'flow' (i.e. are imported or exported) from one country to another. Ma-jor developed economies were found to be typically net importers of embodied carbon emissions — with UK consumption emissions 34% higher than production emissions, and Germany (29%), Japan (19%) and the USA (13%) also significant net importers of embodied emissions.

Effect of Policy

Governments have taken action to reduce GHG emissions (climate change mitigation). Assessments of policy effectiveness have included work by the Intergovernmental Pan-el on Climate Change, International Energy Agency, and United Nations Environment Programme. Policies implemented by governments have included national and region-al targets to reduce emissions, promoting energy efficiency, and support for renewable energy such as Solar energy as an effective use of renewable energy because solar uses energy from the sun and does not release pollutants into the air.

Countries and regions listed in Annex I of the United Nations Framework Convention on Climate Change (UNFCCC) (i.e., the OECD and former planned economies of the Soviet Union) are required to submit periodic assessments to the UNFCCC of actions they are taking to address climate change. Analysis by the UNFCCC (2011) suggested that policies and measures undertaken by Annex I Parties may have produced emis-sion savings of 1.5 thousand Tg CO_2-eq in the year 2010, with most savings made in the energy sector. The projected emissions saving of 1.5 thousand Tg CO_2-eq is measured against a hypothetical "baseline" of Annex I emissions, i.e., projected Annex I emis-sions in the absence of policies and measures. The total projected Annex I saving of 1.5 thousand CO_2-eq does not include emissions savings in seven of the Annex I Parties.

Projections

A wide range of projections of future GHG emissions have been produced. Rogner *et al.* (2007) assessed the scientific literature on GHG projections. Rogner *et al.* (2007) concluded that unless energy policies changed substantially, the world would continue to depend on fossil fuels until 2025–2030. Projections suggest that more than 80% of the world's energy will come from fossil fuels. This conclusion was based on "much evidence" and "high agreement" in the literature. Projected annual energy-related CO_2 emissions in 2030 were 40–110% higher than in 2000, with two-thirds of the increase originating in developing countries. Projected annual per capita emissions in developed country regions remained substantially lower (2.8–5.1 tonnes CO_2) than those in developed country regions (9.6–15.1 tonnes CO_2). Projections consistently showed increase in annual world GHG emissions (the "Kyoto" gases, measured in CO_2-equivalent) of 25–90% by 2030, compared to 2000.

Relative CO_2 Emission from Various Fuels

One liter of gasoline, when used as a fuel, produces 2.32 kg (about 1300 liters or 1.3 cubic meters) of carbon dioxide, a greenhouse gas. One US gallon produces 19.4 lb (1,291.5 gallons or 172.65 cubic feet)

Mass of carbon dioxide emitted per quantity of energy for various fuels			
Fuel name	CO_2 emitted (lbs/10^6 Btu)	CO_2 emitted (g/MJ)	CO_2 emitted (g/KWh)
Natural gas	117	50.30	181.08
Liquefied petroleum gas	139	59.76	215.14
Propane	139	59.76	215.14
Aviation gasoline	153	65.78	236.81
Automobile gasoline	156	67.07	241.45
Kerosene	159	68.36	246.10
Fuel oil	161	69.22	249.19
Tires/tire derived fuel	189	81.26	292.54
Wood and wood waste	195	83.83	301.79
Coal (bituminous)	205	88.13	317.27
Coal (sub-bituminous)	213	91.57	329.65
Coal (lignite)	215	92.43	332.75
Petroleum coke	225	96.73	348.23
Tar-sand Bitumen			
Coal (anthracite)	227	97.59	351.32

Life-cycle Greenhouse-gas Emissions of Energy Sources

A literature review of numerous energy sources CO_2 emissions by the IPCC in 2011, found that, the CO_2 emission value that fell within the 50th percentile of all total life cycle emissions studies conducted was as follows.

Lifecycle greenhouse gas emissions by electricity source.		
Technology	Description	50th percentile (g CO_2/kWh$_e$)
Hydroelectric	reservoir	4
Ocean Energy	wave and tidal	8
Wind	onshore	12
Nuclear	various generation II reactor types	16
Biomass	various	18
Solar thermal	parabolic trough	22
Geothermal	hot dry rock	45
Solar PV	Polycrystalline silicon	46
Natural gas	various combined cycle turbines without scrubbing	469
Coal	various generator types without scrubbing	1001

Removal from the Atmosphere ("Sinks")

Natural Processes

Greenhouse gases can be removed from the atmosphere by various processes, as a consequence of:

- a physical change (condensation and precipitation remove water vapor from the atmosphere).

- a chemical reaction within the atmosphere. For example, methane is oxidized by reaction with naturally occurring hydroxyl radical, OH· and degraded to CO_2 and water vapor (CO_2 from the oxidation of methane is not included in the methane Global warming potential). Other chemical reactions include solution and solid phase chemistry occurring in atmospheric aerosols.

- a physical exchange between the atmosphere and the other compartments of the planet. An example is the mixing of atmospheric gases into the oceans.

- a chemical change at the interface between the atmosphere and the other compartments of the planet. This is the case for CO_2, which is reduced by photosynthesis of plants, and which, after dissolving in the oceans, reacts to form carbonic acid and bicarbonate and carbonatetions.

- a photochemical change. Halocarbons are dissociated by UV light releasing Cl· and F· as free radicals in the stratosphere with harmful effects on ozone (halocarbons are generally too stable to disappear by chemical reaction in the atmosphere).

Negative Emissions

A number of technologies remove greenhouse gases emissions from the atmosphere. Most widely analysed are those that remove carbon dioxide from the atmosphere, either to geologic formations such as bio-energy with carbon capture and storage and carbon dioxide air capture, or to the soil as in the case with biochar. The IPCC has pointed out that many long-term climate scenario models require large scale manmade negative emissions to avoid serious climate change.

History of Scientific Research

In the late 19th century scientists experimentally discovered that N2 and O2 do not absorb infrared radiation (called, at that time, "dark radiation"), while water (both as true vapor and condensed in the form of microscopic droplets suspended in clouds) and CO_2 and other poly-atomic gaseous molecules do absorb infrared radiation. In the early 20th century researchers realized that greenhouse gases in the atmosphere made Earth's overall temperature higher than it would be without them. During the late 20th century, a scientific consensus evolved that increasing concentrations of greenhouse gases in the atmosphere cause a substantial rise in global temperatures and changes to other parts of the climate system, with consequences for the environment and for human health.

Acid Rain

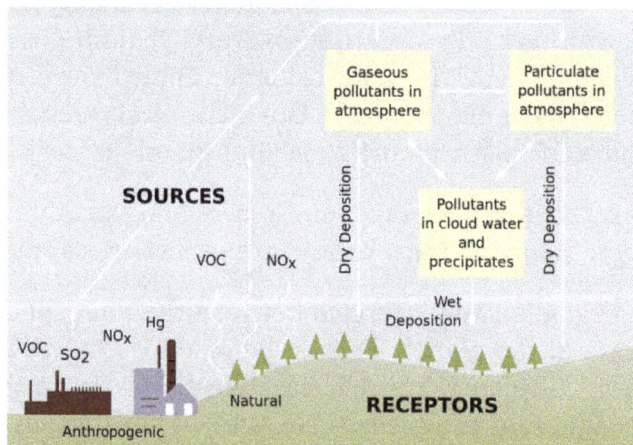

Processes involved in acid deposition (note that only SO_2 and NO_x play a significant role in acid rain).

Acid rain is a rain or any other form of precipitation that is unusually acidic, meaning that it possesses elevated levels of hydrogen ions (low pH). It can have harmful effects on plants, aquatic animals and infrastructure. Acid rain is caused by emissions of sulfur dioxide and nitrogen oxide, which react with the water molecules in the atmosphere to produce acids. Some governments have made efforts since the 1970s to reduce the release of sulfur dioxide and nitrogen oxide into the atmosphere with positive results. Nitrogen oxides can also be produced naturally by lightning strikes, and sulfur dioxide is produced by volcanic eruptions. The chemicals in acid rain can cause paint to peel, corrosion of steel structures such as bridges, and weathering of stone buildings and statues.

Acid clouds can grow on SO_2 emissions from refineries, as seen here in Curaçao.

Definition

"Acid rain" is a popular term referring to the deposition of a mixture from wet (rain, snow, sleet, fog, cloudwater, and dew) and dry (acidifying particles and gases) acidic components. Distilled water, once carbon dioxide is removed, has a neutral pH of 7. Liquids with a pH less than 7 are acidic, and those with a pH greater than 7 are alkaline. "Clean" or unpolluted rain has an acidic pH, but usually no lower than 5.7, because carbon dioxide and water in the air react together to form carbonic acid, a weak acid according to the following reaction:

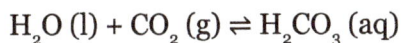

$$H_2O \ (l) + CO_2 \ (g) \rightleftharpoons H_2CO_3 \ (aq)$$

Carbonic acid then can ionize in water forming low concentrations of hydronium and carbonate ions:

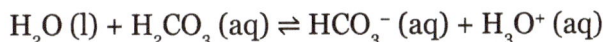

$$H_2O \ (l) + H_2CO_3 \ (aq) \rightleftharpoons HCO_3^- \ (aq) + H_3O^+ \ (aq)$$

However, unpolluted rain can also contain other chemicals which affect its pH (acidity level). A common example is nitric acid produced by electric discharge in the atmosphere such as lightning. Acid deposition as an environmental issue (discussed later in the article) would include additional acids other than H_2CO_3.

History

The corrosive effect of polluted, acidic city air on limestone and marble was noted in the 17th century by John Evelyn, who remarked upon the poor condition of the Arundel marbles. Since the Industrial Revolution, emissions of sulfur dioxide and nitrogen oxides into the atmosphere have increased. In 1852, Robert Angus Smith was the first to show the relationship between acid rain and atmospheric pollution in Manchester, England.

Though acidic rain was discovered in 1853, it was not until the late 1960s that scientists began widely observing and studying the phenomenon. The term "acid rain" was coined in 1872 by Robert Angus Smith. Canadian Harold Harvey was among the first to research a "dead" lake. Public awareness of acid rain in the U.S increased in the 1970s after The New York Times published reports from the Hubbard Brook Experimental Forest in New Hampshire of the myriad deleterious environmental effects shown to result from it.

Occasional pH readings in rain and fog water of well below 2.4 have been reported in industrialized areas. Industrial acid rain is a substantial problem in China and Russia and areas downwind from them. These areas all burn sulfur-containing coal to generate heat and electricity.

The problem of acid rain has not only increased with population and industrial growth, but has become more widespread. The use of tall smokestacks to reduce local pollution has contributed to the spread of acid rain by releasing gases into regional atmospheric circulation. Often deposition occurs a considerable distance downwind of the emissions, with mountainous regions tending to receive the greatest deposition (simply because of their higher rainfall). An example of this effect is the low pH of rain which falls in Scandinavia.

History of Acid Rain in the United States

Since 1998, Harvard University wraps some of the bronze and marble statues on its campus, such as this "Chinese stele", with waterproof covers every winter, in order to protect them from erosion caused by acid rain and acid snow

In 1980, the U.S. Congress passed an Acid Deposition Act. This Act established an 18-year assessment and research program under the direction of the National Acidic Precipitation Assessment Program (NAPAP). NAPAP looked at the entire problem from a scientific perspective. It enlarged a network of monitoring sites to determine how acidic the precipitation actually was, and to determine long-term trends, and established a network for dry deposition. It looked at the effects of acid rain and funded research on the effects of acid precipitation on freshwater and terrestrial ecosystems, historical buildings, monuments, and building materials. It also funded extensive studies on atmospheric processes and potential control programs.

From the start, policy advocates from all sides attempted to influence NAPAP activities to support their particular policy advocacy efforts, or to disparage those of their opponents. For the U.S. Government's scientific enterprise, a significant impact of NAPAP were lessons learned in the assessment process and in environmental research management to a relatively large group of scientists, program managers and the public.

In 1991, DENR provided its first assessment of acid rain in the United States. It reported that 5% of New England Lakes were acidic, with sulfates being the most common problem. They noted that 2% of the lakes could no longer support Brook Trout, and 6% of the lakes were unsuitable for the survival of many species of minnow. Subsequent Reports to Congress have documented chemical changes in soil and freshwater ecosystems, nitrogen saturation, decreases in amounts of nutrients in soil, episodic acidification, regional haze, and damage to historical monuments.

Meanwhile, in 1989, the U.S. Congress passed a series of amendments to the Clean Air Act. Title IV of these amendments established the Acid Rain Program, a cap and trade system designed to control emissions of sulfur dioxide and nitrogen oxides. Title IV called for a total reduction of about 10 million tons of SO_2 emissions from power plants. It was implemented in two phases. Phase I began in 1995, and limited sulfur dioxide emissions from 110 of the largest power plants to a combined total of 8.7 million tons of sulfur dioxide. One power plant in New England (Merrimack) was in Phase I. Four other plants (Newington, Mount Tom, Brayton Point, and Salem Harbor) were added under other provisions of the program. Phase II began in 2000, and affects most of the power plants in the country.

During the 1990s, research continued. On March 10, 2005, EPA issued the Clean Air Interstate Rule (CAIR). This rule provides states with a solution to the problem of power plant pollution that drifts from one state to another. CAIR will permanently cap emissions of SO_2 and NO_x in the eastern United States. When fully implemented, CAIR will reduce SO_2 emissions in 28 eastern states and the District of Columbia by over 70% and NO_x emissions by over 60% from 2003 levels.

Overall, the program's cap and trade program has been successful in achieving its goals. Since the 1990s, SO_2 emissions have dropped 40%, and according to the Pacific Research Institute, acid rain levels have dropped 65% since 1976. Conventional regulation was used in the European Union, which saw a decrease of over 70% in SO_2 emissions during the same time period.

In 2007, total SO_2 emissions were 8.9 million tons, achieving the program's long-term goal ahead of the 2010 statutory deadline.

In 2007 the EPA estimated that by 2010, the overall costs of complying with the program for businesses and consumers would be $1 billion to $2 billion a year, only one fourth of what was originally predicted. Forbes says: *In 2010, by which time the cap and trade system had been augmented by the George W. Bush administration's Clean Air Interstate Rule, SO2 emissions had fallen to 5.1 million tons.*

Emissions of Chemicals Leading to Acidification

The most important gas which leads to acidification is sulfur dioxide. Emissions of nitrogen oxides which are oxidized to form nitric acid are of increasing importance due to stricter controls on emissions of sulfur containing compounds. 70 Tg(S) per year in the form of SO_2 comes from fossil fuel combustion and industry, 2.8 Tg(S) from wildfires and 7–8 Tg(S) per year from volcanoes.

Natural Phenomena

The principal natural phenomena that contribute acid-producing gases to the atmosphere are emissions from volcanoes. Thus, for example, fumaroles from the Laguna Caliente crater of Poás Volcano create extremely high amounts of acid rain and fog, with acidity as high as a pH of 2, clearing an area of any vegetation and frequently causing irritation to the eyes and lungs of inhabitants in nearby settlements. Acid-producing gasses are also created by biological processes that occur on the land, in wetlands, and in the oceans. The major biological source of sulfur containing compounds is dimethyl sulfide.

Nitric acid in rainwater is an important source of fixed nitrogen for plant life, and is also produced by electrical activity in the atmosphere such as lightning.

Acidic deposits have been detected in glacial ice thousands of years old in remote parts of the globe.

Soils of coniferous forests are naturally very acidic due to the shedding of needles, and the results of this phenomenon should not be confused with acid rain.

Human Activity

The principal cause of acid rain is sulfur and nitrogen compounds from human sources, such as electricity generation, factories, and motor vehicles. Electrical power generation using coal is among the greatest contributors to gaseous pollutions that are responsible for acidic rain. The gases can be carried hundreds of kilometers in the atmosphere before they are converted to acids and deposited. In the past, factories had short funnels to let out smoke but this caused many problems locally; thus, factories now have taller smoke funnels. However, dispersal from these taller stacks causes pollutants to be carried farther, causing widespread ecological damage.

The coal-fired Gavin Power Plant in Cheshire, Ohio

Chemical Processes

Combustion of fuels produces sulfur dioxide and nitric oxides. They are converted into sulfuric acid and nitric acid.

Gas Phase Chemistry

In the gas phase sulfur dioxide is oxidized by reaction with the hydroxyl radical via an intermolecular reaction:

$$SO_2 + OH\cdot \rightarrow HOSO_2\cdot$$

which is followed by:

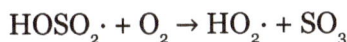

$$HOSO_2\cdot + O_2 \rightarrow HO_2\cdot + SO_3$$

In the presence of water, sulfur trioxide (SO_3) is converted rapidly to sulfuric acid:

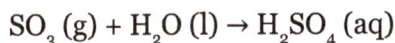

$$SO_3 \ (g) + H_2O \ (l) \rightarrow H_2SO_4 \ (aq)$$

Nitrogen dioxide reacts with OH to form nitric acid:

$$NO_2 + OH\cdot \rightarrow HNO_3$$

Chemistry in Cloud Droplets

When clouds are present, the loss rate of SO_2 is faster than can be explained by gas phase chemistry alone. This is due to reactions in the liquid water droplets.

Hydrolysis

Sulfur dioxide dissolves in water and then, like carbon dioxide, hydrolyses in a series of equilibrium reactions:

$$SO_2\ (g) + H_2O \rightleftharpoons SO_2 \cdot H_2O$$

$$SO_2 \cdot H_2O \rightleftharpoons H^+ + HSO_3^-$$

$$HSO_3^- \rightleftharpoons H^+ + SO_3^{2-}$$

Oxidation

There are a large number of aqueous reactions that oxidize sulfur from S(IV) to S(VI), leading to the formation of sulfuric acid. The most important oxidation reactions are with ozone, hydrogen peroxide and oxygen (reactions with oxygen are catalyzed by iron and manganese in the cloud droplets).

Acid Deposition

Wet Deposition

Wet deposition of acids occurs when any form of precipitation (rain, snow, and so on.) removes acids from the atmosphere and delivers it to the Earth's surface. This can result from the deposition of acids produced in the raindrops or by the precipitation removing the acids either in clouds or below clouds. Wet removal of both gases and aerosols are both of importance for wet deposition.

Dry Deposition

Acid deposition also occurs via dry deposition in the absence of precipitation. This can be responsible for as much as 20 to 60% of total acid deposition. This occurs when particles and gases stick to the ground, plants or other surfaces.

Adverse Effects

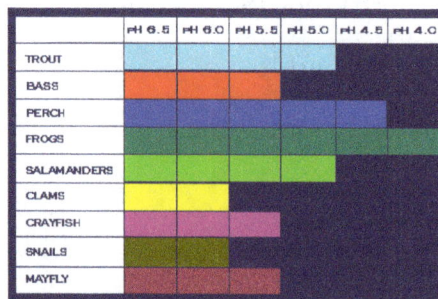

	pH 6.5	pH 6.0	pH 5.5	pH 5.0	pH 4.5	pH 4.0
TROUT						
BASS						
PERCH						
FROGS						
SALAMANDERS						
CLAMS						
CRAYFISH						
SNAILS						
MAYFLY						

This chart shows that not all fish, shellfish, or the insects that they eat can tolerate the same amount of acid; for example, frogs can tolerate water that is more acidic (i.e., has a lower pH) than trout.

Acid rain has been shown to have adverse impacts on forests, freshwaters and soils, killing insect and aquatic life-forms as well as causing damage to buildings and having impacts on human health.

Surface Waters and Aquatic Animals

Both the lower pH and higher aluminium concentrations in surface water that occur as a result of acid rain can cause damage to fish and other aquatic animals. At pHs lower than 5 most fish eggs will not hatch and lower pHs can kill adult fish. As lakes and rivers become more acidic biodiversity is reduced. Acid rain has eliminated insect life and some fish species, including the brook trout in some lakes, streams, and creeks in geographically sensitive areas, such as the Adirondack Mountains of the United States. However, the extent to which acid rain contributes directly or indirectly via runoff from the catchment to lake and river acidity (i.e., depending on characteristics of the surrounding watershed) is variable. The United States Environmental Protection Agency's (EPA) website states: "Of the lakes and streams surveyed, acid rain caused acidity in 75% of the acidic lakes and about 50% of the acidic streams".

Soils

Soil biology and chemistry can be seriously damaged by acid rain. Some microbes are unable to tolerate changes to low pH and are killed. The enzymes of these microbes are denatured (changed in shape so they no longer function) by the acid. The hydronium ions of acid rain also mobilize toxins such as aluminium, and leach away essential nutrients and minerals such as magnesium.

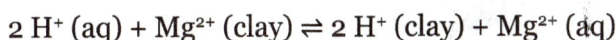

$$2 H^+ (aq) + Mg^{2+} (clay) \rightleftharpoons 2 H^+ (clay) + Mg^{2+} (aq)$$

Soil chemistry can be dramatically changed when base cations, such as calcium and magnesium, are leached by acid rain thereby affecting sensitive species, such as sugar maple (Acer saccharum).

Forests and Other Vegetation

Adverse effects may be indirectly related to acid rain, like the acid's effects on soil or high concentration of gaseous precursors to acid rain. High altitude forests are especially vulnerable as they are often surrounded by clouds and fog which are more acidic than rain.

Other plants can also be damaged by acid rain, but the effect on food crops is minimized by the application of lime and fertilizers to replace lost nutrients. In cultivated areas, limestone may also be added to increase the ability of the soil to keep the pH stable, but this tactic is largely unusable in the case of wilderness lands. When calcium is leached from the needles of red spruce, these trees become less cold tolerant and exhibit winter injury and even death.

Ocean Acidification

Coral's limestone skeletal is sensitive to pH drop, because the calcium carbonate, core component of the limestone dissolves in acidic (low pH) solutions.

Human Health Effects

Acid rain does not directly affect human health. The acid in the rainwater is too dilute to have direct adverse effects. However, the particulates responsible for acid rain (sulfur dioxide and nitrogen oxides) do have an adverse effect. Increased amounts of fine particulate matter in the air do contribute to heart and lung problems including asthma and bronchitis.

Other Adverse Effects

Effect of acid rain on statues

Acid rain and weathering

Acid rain can damage buildings, historic monuments, and statues, especially those made of rocks, such as limestone and marble, that contain large amounts of calcium carbonate. Acids in the rain react with the calcium compounds in the stones to create gypsum, which then flakes off.

$$CaCO_3 \text{ (s)} + H_2SO_4 \text{ (aq)} \rightleftharpoons CaSO_4 \text{ (s)} + CO_2 \text{ (g)} + H_2O \text{ (l)}$$

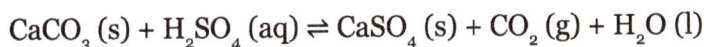

The effects of this are commonly seen on old gravestones, where acid rain can cause the inscriptions to become completely illegible. Acid rain also increases the corrosion rate of metals, in particular iron, steel, copper and bronze.

Affected Areas

Places significantly impacted by acid rain around the globe include most of eastern Europe from Poland northward into Scandinavia, the eastern third of the United States, and southeastern Canada. Other affected areas include the southeastern coast of China and Taiwan.

Prevention Methods

Technical Solutions

Many coal-firing power stations use flue-gas desulfurization (FGD) to remove sulfur-containing gases from their stack gases. For a typical coal-fired power station, FGD will remove 95% or more of the SO_2 in the flue gases. An example of FGD is the wet scrubber which is commonly used. A wet scrubber is basically a reaction tower equipped with a fan that extracts hot smoke stack gases from a power plant into the tower. Lime or limestone in slurry form is also injected into the tower to mix with the stack gases and combine with the sulfur dioxide present. The calcium carbonate of the limestone produces pH-neutral calcium sulfate that is physically removed from the scrubber. That is, the scrubber turns sulfur pollution into industrial sulfates.

In some areas the sulfates are sold to chemical companies as gypsum when the purity of calcium sulfate is high. In others, they are placed in landfill. However, the effects of acid rain can last for generations, as the effects of pH level change can stimulate the continued leaching of undesirable chemicals into otherwise pristine water sources, killing off vulnerable insect and fish species and blocking efforts to restore native life.

Fluidized bed combustion also reduces the amount of sulfur emitted by power production.

Vehicle emissions control reduces emissions of nitrogen oxides from motor vehicles.

International Treaties

A number of international treaties on the long-range transport of atmospheric pollutants have been agreed for example, the 1985 Helsinki Protocol on the Reduction of Sulphur Emissions under the Convention on Long-Range Transboundary Air Pollution. Canada and the US signed the Air Quality Agreement in 1991. Most European countries and Canada have signed the treaties.

Emissions Trading

In this regulatory scheme, every current polluting facility is given or may purchase on an open market an emissions allowance for each unit of a designated pollutant it emits. Operators can then install pollution control equipment, and sell portions of their emissions allowances they no longer need for their own operations, thereby recovering some of the capital cost of their investment in such equipment. The intention is to give operators economic incentives to install pollution controls.

The first emissions trading market was established in the United States by enactment of the Clean Air Act Amendments of 1990. The overall goal of the Acid Rain Program established by the Act is to achieve significant environmental and public health benefits through reductions in emissions of sulfur dioxide (SO_2) and nitrogen oxides (NO_x), the primary causes of acid rain. To achieve this goal at the lowest cost to society, the program employs both regulatory and market based approaches for controlling air pollution.

Drug Pollution

Drug pollution or pharmaceutical pollution is pollution of the environment with pharmaceutical drugs and their metabolites, which reach the marine environment (groundwater, rivers, lakes, and oceans) through wastewater. Drug pollution is therefore mainly a form of water pollution.

"Pharmaceutical pollution is now detected in waters throughout the world," said a scientist at the Cary Institute of Ecosystem Studies in Millbrook, New York. "Causes include aging infrastructure, sewage overflows and agricultural runoff. Even when wastewater makes it to sewage treatment facilities, they aren't equipped to remove pharmaceuticals."

Sources and Effects

Most such pollution comes simply from the drugs having been cleared and excreted in the urine. The portion that comes from expired or unneeded drugs that are flushed unused down the toilet is smaller, but it is also important, especially in hospitals (where its magnitude is greater than in residential contexts). Other sources include agricultural runoff (because of antibiotic use in livestock) and pharmaceutical manufacturing. Drug pollution is implicated in the sex effects of water pollution. It is suspected as a contributor (besides industrial pollution) in fish kills, amphibian dieoffs, and amphibian pathomorphology.

Prevention

The main action for preventing drug pollution is to incinerate unwanted pharmaceutical drugs rather than flushing them down the drain. Burning them chemically degrades their active molecules, with few exceptions. The resulting ash can be further processed before landfilling, such as to remove and recycle any heavy metals that may be present.

There are now programs in many cities that provide collection points at places including drug stores, grocery stores, and police stations. People can bring their unwanted pharmaceuticals there for safe disposal, instead of flushing them (externalizing them to the waterways) or throwing them in the trash (externalizing them to the landfill, where they can become leachate).

Another aspect of drug pollution prevention is environmental law and regulation, although this faces problems of enforcement costs, enforcement corruption and negligence, and, where enforcement succeeds, increased costs of doing business. The lobbying of pros and cons is ongoing.

Manufacturing

One extreme example of drug pollution was found in India in 2009 in an area where pharmaceutical manufacturing activity is concentrated. Not all pharmaceutical manufacturing contributes to the problem. In places where environmental law and regulation are adequately enforced, the wastewater from the factories is cleaned to a safe level. But to the extent that the market rewards "looking the other way" in developing nations, whether through local corruption (bribed inspectors or regulators) or plausible deniability, such protections are circumvented. This problem belongs to everyone, because consumers in well-regulated places constitute the biggest customers of the factories that operate in the inadequately regulated or inspected places, meaning that externality is involved.

Chemical Waste

Chemical waste is a waste that is made from harmful chemicals (mostly produced by large factories). Chemical waste may fall under regulations such as COSHH in the United Kingdom, or the Clean Water Act and Resource Conservation and Recovery Act in the United States. In the U.S., the Environmental Protection Agency (EPA) and the Occupational Safety and Health Administration (OSHA), as well as state and local regulations also regulate chemical use and disposal. Chemical waste may or may not be classed as hazardous waste. A chemical hazardous waste is a solid, liquid, or gaseous material that displays either a "Hazardous Characteristic" or is specifically "listed" by name as a hazardous waste. There are four characteristics chemical wastes may have to be considered as hazardous. These are Ignitability, Corrosivity, Reactivity, and Toxicity. This type of hazardous waste must be categorized as to its identity, constituents, and hazards so that it may be safely handled and managed. Chemical waste is a broad term and encompasses many types of materials. Consult the Material Safety Data Sheet (MSDS), Product Data Sheet or Label for a list of constituents. These sources should state weather this chemical waste is a waste that needs special disposal.

Chemical Waste Bin (Chemobox)

Guidance for Disposal of Laboratory Chemical Wastes

In the laboratory, chemical wastes are usually segregated on-site into appropriate waste carboys, and disposed by a specialist contractor in order to meet safety, health, and legislative requirements.

Chemical waste category that should be followed for proper packaging, labelling, and disposal of chemical waste.

Innocuous aqueous waste (such as solutions of sodium chloride) may be poured down the sink. Some chemicals are washed down with excess water. This includes: concentrated and dilute acids and alkalis, harmless soluble inorganic salts (all drying agents), alcohols containing salts, hypochlorite solutions, fine (tlc grade) silica and alumina. Aqueous waste containing toxic compounds are collected separately

Waste elemental mercury, spent acids and bases may be collected separately for recycling.

Waste organic solvents are separated into chlorinated and non-chlorinated solvent waste. Chlorinated solvent waste is usually incinerated at high temperature to minimize the formation of dioxins. Non-chlorinated solvent waste can be burned for energy recovery.

In contrast to this, chemical materials on the "Red List" should never be washed down a drain. This list includes: compounds with transitional metals, biocides, cyanides, mineral oils and hydrocarbons, poisonous organosilicon compounds, metal phosphides, phosphorus element, and fluorides and nitrites.

Moreover, the Environmental Protection Agency (EPA) prohibits disposing certain materials down any UVM drain. Including flammable liquids, liquids capable of causing damage to wastewater facilities (this can be determined by the pH), highly viscous materials capable of causing an obstruction in the wastewater system, radioactive materials, materials that have or create a strong odor, wastewater capable of significantly raising the temperature of the system, and pharmaceuticals or endocrine disruptors.

Broken glassware are usually collected in plastic-lined cardboard boxes for landfilling. Due to contamination, they are usually not suitable for recycling. Similarly, used hypodermic needles are collected as sharps and are incinerated as medical waste.

Chemical Compatibility Guideline

Many chemicals may react adversely when combined. It's recommended that incompatible chemicals are stored in separate areas of the lab.

Acids should be separated from alkalis, metals, cyanides, sulfides, azides, phosphides, and oxidizers. The reason being, when combined acids with these type of compounds, violent exothermic reaction can occur possibly causing flammable gas, and in some cases explosions.

Oxidizers should be separated from acids, organic materials, metals, reducing agents, and ammonia. This is because when combined oxidizers with these type of compounds, inflammable, and sometimes toxic compounds can occur.

Container Compatibility

When disposing hazardous laboratory chemical waste, chemical compatibility must be considered. For safe disposal, the container must be chemically compatible with the material it will hold. Chemicals must not react with, weaken, or dissolve the container or lid. Acids or bases should not be stored in metal. Hydrofluoric acid should not store in glass. Gasoline (solvents) should not store or transport in lightweight polyethylene containers such as milk jugs. Moreover, the Chemical Compatibility Guidelines should be considered for more detailed information.

Laboratory Waste Containers

Packaging, labelling, storage are the three requirements for disposing chemical waste.

Packaging

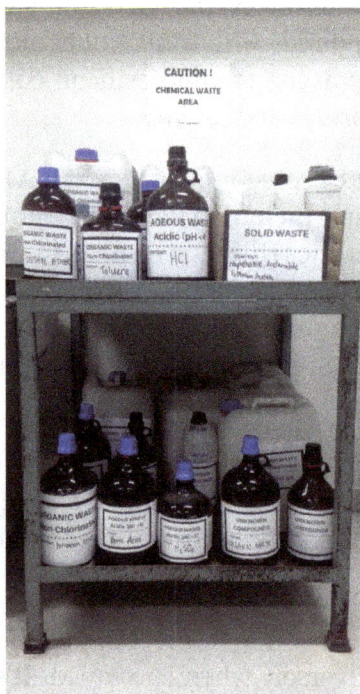

How to properly label, package, and store chemical waste safely.

For packaging, chemical liquid waste containers should only be filled up to 75% capacity to allow for vapour expansion and to reduce potential spills which could occur from moving overfilled containers. Container material must be compatible with the stored hazardous waste. Finally, wastes must not be packaged in containers that improperly identify other nonexisting hazards.

In addition to the general packaging requirements mentioned above, incompatible materials should never be mixed together in a single container. Wastes must be stored in containers compatible with the chemicals stored as mentioned in the container compatibility section. Solvent safety cans should to be used to collect and temporarily store large volumes (10-20 litres) of flammable organic waste solvents, precipitates, solids or other non-fluid wastes should not be mixed into safety cans.

Labelling

Label all containers with the group name from the chemical waste category and an itemized list of the contents. All chemicals or anything contaminated with chemicals posing a significant hazard. All waste must be appropriately packaged.

Storage

When storing chemical wastes, the containers must be in good condition and should remain closed unless waste is being added. Hazardous waste must be stored safely prior to removal from the laboratory and should not be allowed to accumulate. Container should be sturdy and leakproof, also has to be labeled. All liquid waste must be stored in leakproof containers with a screw- top or other secure lid. Snap caps, mis-sized caps, parafilm and other loose fitting lids are not acceptable. If necessary, transfer waste material to a container that can be securely closed. Keep waste containers closed except when adding waste. Secondary containment should be in place to capture spills and leaks from the primary container, segregate incompatible hazardous wastes, such as acids and bases.

Mapping of Chemical Waste in the United States

TOXMAP is a Geographic Information System (GIS) from the Division of Specialized Information Services of the United States National Library of Medicine (NLM) that uses maps of the United States to help users visually explore data from the United States Environmental Protection Agency's (EPA) Toxics Release Inventory and Superfund Basic Research Programs. TOXMAP is a resource funded by the US Federal Government. TOXMAP's chemical and environmental health information is taken from NLM's Toxicology Data Network (TOXNET) and PubMed, and from other authoritative sources.

Chemical Waste in Canadian Aquaculture

Green Sea Urchin or S. droebacheinsis

Chemical waste in our oceans is becoming a major issue for the marine life. There have been many studies conducted to try and prove the effects of these chemical in our oceans. In Canada, many of the studies concentrated on the Atlantic provinces, where fishing and aquaculture are an important part of the economy. In New Brunswick, a study was done on the sea urchin in an attempt to identify the effects of toxic and chemical waste on life beneath the ocean, specifically the wasted from the salmon farms. Sea urchins were used to check the levels of metals in the environment. It is advantageous to use green sea urchins, Strongylocentrotus droebachiensis, because they are widely distributed, abundant in many locations, and easily accessible. By investigating the concentrations of metals in the green sea urchins, the impacts of produced chemicals from salmon aquaculture activity could be assessed and detected. Samples were taken at 25m intervals along a transect in the direction of the main tidal flow. The study found that there was impacts to at least 75m based on the intestine metal concentrations. So based on this study it is clear that the metals are contaminating the oceans and negatively affecting aquatic life.

Uranium in Ground and Surface Water in Canada

Another issue regarding chemical waste is the potential risk of surface and groundwater contamination by the heavy metals and radionuclides leached from uranium waste-rock piles (UWRP) A Radionuclide is an atom that has excess nuclear energy, making it unstable. Uranium waste-rock piles refers to Uranium mining, which is the process of extraction of uranium ore from the ground. . An example of such threats is in Saskatchewan, Uranium mining and ore processing (milling) can pose a threat to the environment. In open pit mining, large amounts of materials are excavated and disposed off in waste-rock piles. Waste-rock piles from the Uranium mining industry can contain several heavy metals and contaminants that may become mobile under certain conditions. Environmental contaminants may include acid mine drainage, higher concentrations of radionuclides, and non-radioactive metals/metalloids (i.e. As, Mo, Ni, Cu, Zn).

The leachability of heavy metals and radionuclide from UWRP plays a significant role in determining their potential environmental risks to surrounding surface and groundwater. Substantial differences in the solid-phase partitioning and chemical leachability of Ni and U were observed in the investigated UWRP lithological materials and background organic-rich lake sediment. For Instance, in the uranium-mining district of Northern Saskatchewan, Canada, the sequential extraction results showed that a significant amount of Ni (Nickel) was present in the non-labile residual fraction, while Uranium was mostly distributed in the moderately labile fractions. Although Nickel was much less labile than Uranium, the observed Nickel exceeded Uranium concentrations in leaching]].The observed Nickel and Uranium concentrations were relatively high in the underlying organic-rich lake sediment. Expressed as the percentage of total metal content, potential leachability decreased in the order U > Ni. Data suggest that these elements could potentially migrate to the water table below the UWRP. Detailed information regarding the solid-phase distribution of contaminants in the UWRP is critical to understand the potential for their environmental transport and mobility

The most visible civilian use of uranium is as the thermal power source used in nuclear power plants

References

- Davis, Devra (2002). When Smoke Ran Like Water: Tales of Environmental Deception and the Battle Against Pollution. Basic Books. ISBN 0-465-01521-2.

- Turner, D.B. (1994). Workbook of atmospheric dispersion estimates: an introduction to dispersion modeling (2nd ed.). CRC Press. ISBN 1-56670-023-X.

- Wallace, John M. and Peter V. Hobbs. Atmospheric Science; An Introductory Survey.Elsevier. Second Edition, 2006. ISBN 978-0-12-732951-2.

- Evans, Kimberly Masters (2005). "The greenhouse effect and climate change". The environment: a revolution in attitudes. Detroit: Thomson Gale. ISBN 0-7876-9082-1.

- World Energy Outlook 2009 (PDF), Paris, France: International Energy Agency (IEA), 2009, pp. 179–180, ISBN 978-92-64-06130-9

- Berresheim, H.; Wine, P.H. and Davies D.D. (1995). "Sulfur in the Atmosphere". In Composition, Chemistry and Climate of the Atmosphere, ed. H.B. Singh. Van Nostrand Rheingold ISBN 0-442-01264-0

- DeHayes, D.H., Schaberg, P.G. and G.R. Strimbeck. (2001). Red Spruce Hardiness and Freezing Injury Susceptibility. In: F. Bigras, ed. Conifer Cold Hardiness. Kluwer Academic Publishers, the Netherlands ISBN 0-7923-6636-0.

- World energy outlook 2007 edition – China and India insights. International Energy Agency (IEA), Head of Communication and Information Office, 9 rue de la Fédération, 75739 Paris Cedex 15, France. 2007. p. 600. ISBN 978-92-64-02730-5. Retrieved 2010-05-04.

- Seinfeld, John H.; Pandis, Spyros N (1998). Atmospheric Chemistry and Physics — From Air Pollution to Climate Change. John Wiley and Sons, Inc. ISBN 978-0-471-17816-3

- Hallam, Bill (April–May 2010). "Techniques for Efficient Hazardous Chemicals Handling and Disposal". Pollution Equipment News. p. 13. Retrieved 10 March 2016.

- "LABORATORY CHEMICAL WASTE MANAGEMENT GUIDELINES" (PDF). Environmental Health and Radiation Safety University of Pennsylvania. Retrieved 10 March 2016.

- "Waste - Disposal of Laboratory Wastes (GUIDANCE) | Current Staff | University of St Andrews". www.st-andrews.ac.uk. Retrieved 2016-02-04.

7

Human Impact, Analysis and Assessment

Human activities have greatly affected the environment. This can be described by the formula I = PAT, in words this means the impact humans have on the environment equals to the production of population, affluence and technology. Human impacts on the nitrogen cycle and life cycle assessment have also been explained in the following section. This chapter is a compilation of the various branches of environmental monitoring that form an integral part of the broader subject matter.

Human Impact on the Environment

Human impact on the environment or anthropogenic impact on the environment includes impacts on biophysical environments, biodiversity, and other resources. The term *anthropogenic* designates an effect or object resulting from human activity. The term was first used in the technical sense by Russian geologist Alexey Pavlov, and was first used in English by British ecologist Arthur Tansley in reference to human influences on climax plant communities. The atmospheric scientist Paul Crutzen introduced the term "anthropocene" in the mid-1970s. The term is sometimes used in the context of pollution emissions that are produced as a result of human activities but applies broadly to all major human impacts on the environment.

The ecosystem of public parks often includes humans feeding the wildlife.

Causes

Technology

The applications of technology often result in unavoidable and unexpected environmental impacts, which according to the I = PAT equation is measured as resource use or pollution generated per unit GDP. Environmental impacts caused by the application of technology are often perceived as unavoidable for several reasons. First, given that the purpose of many technologies is to exploit, control, or otherwise "improve" upon nature for the perceived benefit of humanity while at the same time the myriad of processes in nature have been optimized and are continually adjusted by evolution, any disturbance of these natural processes by technology is likely to result in negative environmental consequences. Second, the conservation of mass principle and the first law of thermodynamics (i.e., conservation of energy) dictate that whenever material resources or energy are moved around or manipulated by technology, environmental consequences are inescapable. Third, according to the second law of thermodynamics, order can be increased within a system (such as the human economy) only by increasing disorder or entropy outside the system (i.e., the environment). Thus, technologies can create "order" in the human economy (i.e., order as manifested in buildings, factories, transportation networks, communication systems, etc.) only at the expense of increasing "disorder" in the environment. According to a number of studies, increased entropy is likely to be correlated to negative environmental impacts.

Agriculture

The environmental impact of agriculture varies based on the wide variety of agricultural practices employed around the world. Ultimately, the environmental impact depends on the production practices of the system used by farmers. The connection between emissions into the environment and the farming system is indirect, as it also depends on other climate variables such as rainfall and temperature.

There are two types of indicators of environmental impact: "means-based", which is based on the farmer's production methods, and "effect-based", which is the impact that farming methods have on the farming system or on emissions to the environment. An example of a means-based indicator would be the quality of groundwater, that is effected by the amount of nitrogen applied to the soil. An indicator reflecting the loss of nitrate to groundwater would be effect-based.

The environmental impact of agriculture involves a variety of factors from the soil, to water, the air, animal and soil diversity, plants, and the food itself. Some of the environmental issues that are related to agriculture are climate change, deforestation, genetic engineering, irrigation problems, pollutants, soil degradation, and waste.

Fishing

The environmental impact of fishing can be divided into issues that involve the availability of fish to be caught, such as overfishing, sustainable fisheries, and fisheries management; and issues that involve the impact of fishing on other elements of the environment, such as by-catch and destruction of habitat such as coral reefs.

Fishing down the foodweb.

These conservation issues are part of marine conservation, and are addressed in fisheries science programs. There is a growing gap between how many fish are available to be caught and humanity's desire to catch them, a problem that gets worse as the world population grows.

Similar to other environmental issues, there can be conflict between the fishermen who depend on fishing for their livelihoods and fishery scientists who realize that if future fish populations are to be sustainable then some fisheries must reduce or even close.

The journal *Science* published a four-year study in November 2006, which predicted that, at prevailing trends, the world would run out of wild-caught seafood in 2048. The scientists stated that the decline was a result of overfishing, pollution and other environmental factors that were reducing the population of fisheries at the same time as their ecosystems were being degraded. Yet again the analysis has met criticism as being fundamentally flawed, and many fishery management officials, industry representatives and scientists challenge the findings, although the debate continues. Many countries, such as Tonga, the United States, Australia and New Zealand, and international management bodies have taken steps to appropriately manage marine resources.

Irrigation

The environmental impact of irrigation includes the changes in quantity and quality of soil and water as a result of irrigation and the ensuing effects on natural and social conditions at the tail-end and downstream of the irrigation scheme.

The impacts stem from the changed hydrological conditions owing to the installation and operation of the scheme.

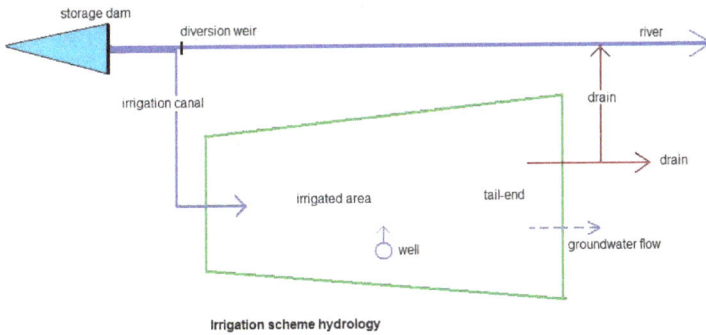

Irrigation scheme hydrology

An irrigation scheme often draws water from the river and distributes it over the irrigated area. As a hydrological result it is found that:

- the downstream river discharge is reduced

- the evaporation in the scheme is increased

- the groundwater recharge in the scheme is increased

- the level of the water table rises

- the drainage flow is increased.

These May be Called Direct Effects.

Effects on soil and water quality are indirect and complex, and subsequent impacts on natural, ecological and socio-economic conditions are intricate. In some, but not all instances, water logging and soil salinization can result. However, irrigation can also be used, together with soil drainage, to overcome soil salinization by leaching excess salts from the vicinity of the root zone.

Irrigation can also be done extracting groundwater by (tube)wells. As a hydrological result it is found that the level of the water descends. The effects may be water mining, land/soil subsidence, and, along the coast, saltwater intrusion.

Irrigation projects can have large benefits, but the negative side effects are often over-looked. Agricultural irrigation technologies such as high powered water pumps, dams, and pipelines are responsible for the large-scale depletion of fresh water resources such as aquifers, lakes, and rivers. As a result of this massive diversion of freshwater, lakes, rivers, and creeks are running dry, severely altering or stressing surrounding ecosys-tems, and contributing to the extinction of many aquatic species.

Agricultural Land Loss and Soil Erosion

Lal and Stewart estimated global loss of agricultural land by degradation and aban-donment at 12 million hectares per year. In contrast, according to Scherr, GLASOD (Global Assessment of Human-Induced Soil Degradation, under the UN Environment

Programme) estimated that 6 million hectares of agricultural land per year had been lost to soil degradation since the mid-1940s, and she noted that this magnitude is similar to earlier estimates by Dudal and by Rozanov et al. Such losses are attributable not only to soil erosion, but also to salinization, loss of nutrients and organic matter, acidification, compaction, water logging and subsidence. Human-induced land degradation tends to be particularly serious in dry regions. Focusing on soil properties, Oldeman estimated that about 19 million square kilometers of global land area had been degraded; Dregne and Chou, who included degradation of vegetation cover as well as soil, estimated about 36 million square kilometers degraded in the world's dry regions. Despite estimated losses of agricultural land, the amount of arable land used in crop production globally increased by about 9 percent from 1961 to 2012, and is estimated to have been 1.396 billion hectares in 2012.

Global average soil erosion rates are thought to be high, and erosion rates on conventional cropland generally exceed estimates of soil production rates, usually by more than an order of magnitude. In the US, sampling for erosion estimates by the US NRCS (Natural Resources Conservation Service) is statistically based, and estimation uses the Universal Soil Loss Equation and Wind Erosion Equation. For 2010, annual average soil loss by sheet, rill and wind erosion on non-federal US land was estimated to be 10.7 t/ha on cropland and 1.9 t/ha on pasture land; the average soil erosion rate on US cropland had been reduced by about 34 percent since 1982. No-till and low-till practices have become increasingly common on North American cropland used for production of grains such as wheat and barley. On uncultivated cropland, the recent average total soil loss has been 2.2 t/ha per year. In comparison with agriculture using conventional cultivation, it has been suggested that, because no-till agriculture produces erosion rates much closer to soil production rates, it could provide a foundation for sustainable agriculture.

Meat Production

Environmental impacts associated with meat production include use of fossil energy, water and land resources, greenhouse gas emissions, and in some instances, rainforest clearing, water pollution and species endangerment, among other adverse effects. Steinfeld et al. of the FAO estimated that 18 percent of global anthropogenic GHG (greenhouse gas) emissions (estimated as 100-year carbon dioxide equivalents) are associated in some way with livestock production. A more recent FAO analysis estimated that all agriculture, including the livestock sector, in 2011 accounted for 12 percent of global anthropogenic GHG emissions expressed as 100-year carbon dioxide equivalents. Similarly, the Intergovernmental Panel on Climate Change has estimated that about 10 to 12 percent of global anthropogenic GHG emissions (expressed as 100-year carbon dioxide equivalents) were assignable to all of agriculture, including the livestock sector, in 2005 and again in 2010. The percentage assignable to livestock would be some fraction of the percentage for agriculture. The amount assignable to meat production would be some fraction of that assigned to livestock. FAO data indicate that meat accounted for 26 percent of global livestock product tonnage in 2011. However, many estimates use different sectoral assignment of some emissions. Environmental specialists Jeff Anhang and Robert Goodland with the IFC and World Bank, have put the GHG associated with livestock at 51%, pointing out the FAO report failed to account for the 8,769 metric tons of respiratory CO_2 produced each year, undercounted methane production and land use associated with livestock, and failed to properly categorize emissions related to the slaughtering, processing, packaging, storing and transporting of animals and animal products.

Globally, enteric fermentation (mostly in ruminant livestock) accounts for about 27 percent of anthropogenic methane emissions, Despite methane's 100-year global warming potential, recently estimated at 28 without and 34 with climate carbon feedbacks, methane emission is currently contributing relatively little to global warming. Over the decade 2000 through 2009, atmospheric methane content increased by an average of only 6 Tg per year (because nearly all natural and anthropogenic methane emission was offset by degradation), while atmospheric carbon dioxide increased by nearly 15,000 Tg per year. At the currently estimated rate of methane degradation, slight reduction of anthropogenic methane emissions, to about 98 percent of that decade's average, would be expected to result in no further increase of atmospheric methane content. Although reduction of methane emissions would have a rapid effect on warming, the expected effect would be small. Other anthropogenic GHG emissions associated with livestock production include carbon dioxide from fossil fuel consumption (mostly for production, harvesting and transport of feed), and nitrous oxide emissions associated with use of nitrogenous fertilizers, growing of nitrogen-fixing legume vegetation and manure management. Management practices that can mitigate GHG emissions from production of livestock and feed have been identified.

Livestock production, including feed production and grazing, uses about 30 percent of the earth's ice-free terrestrial surface: about 26 percent for grazing and about 4 percent for other feed production. The intensity and duration of grazing use vary greatly and these, together with terrain, vegetation and climate, influence the nature and importance of grazing's environmental impact, which can range from severe to negligible, and in some cases (as noted below) beneficial. Excessive use of vegetation by grazing can be especially conducive to land degradation in dry areas.

Considerable water use is associated with meat production, mostly because of water used in production of vegetation that provides feed. There are several published estimates of water use associated with livestock and meat production, but the amount of water use assignable to such production is seldom estimated. For example, "green water" use is evapotranspirational use of soil water that has been provided directly by precipitation; and "green water" has been estimated to account for 94 percent of global beef cattle production's "water footprint", and on rangeland, as much as 99.5 percent of the water use associated with beef production is "green water". However, it would be misleading simply to assign that associated rangeland green water use to beef production, partly because that evapotranspirational use occurs even in the absence of cattle. Even when cattle are present, most of that associated water use can be considered assignable to production of terrestrial environmental values, because it produces root and residue biomass important for erosion control, stabilization of soil structure, nutrient cycling, carbon sequestration, support of numerous primary consumers, many of which support higher trophic levels, etc. Withdrawn water (from surface and groundwater sources) is used for livestock watering, and in some cases is also used for irrigation of forage and feed crops. Whereas all irrigation in the US (including loss in conveyance) is estimated to account for about 38 percent of US withdrawn freshwater use, irrigation water for production of livestock feed and forage has been estimated to account for about 9 percent; other withdrawn freshwater use for the livestock sector (for drinking, washdown of facilities, etc.) is estimated at about 0.7 percent. Because of the preponderance of non-meat products from the livestock sector only some fraction of this water use is assignable to meat production.

Impairment of water quality by manure and other substances in runoff and infiltrating water is a concern, especially where intensive livestock production is carried out. In the US, in a comparison of 32 industries, the livestock industry was found to have a relatively good record of compliance with environmental regulations pursuant to the Clean Water Act and Clean Air Act, but pollution issues from large livestock operations can sometimes be serious where violations occur. Various measures have been suggested by the US Environmental Protection Agency, among others, which can help reduce livestock damage to streamwater quality and riparian environments.

Data of a USDA study indicate that, in 2002, about 0.6 percent of non-solar energy use in the United States was accounted for by production of meat-producing livestock and poultry. This estimate included embodied energy used in production, such as energy used in manufacture and transport of fertilizer for feed production. (Non-solar energy is specified, because solar energy is used in such processes as photosynthesis and hay-drying.)

Changes in livestock production practices influence the environmental impact of meat production, as illustrated by some beef data. In the US beef production system, practices prevailing in 2007 are estimated to have involved 8.6 percent less fossil fuel use, 16.3 percent less greenhouse gas emissions (estimated as 100-year carbon dioxide equivalents), 12.1 percent less withdrawn water use and 33.0 percent less land use, per unit mass of beef produced, than in 1977. From 1980 to 2012 in the US, while population increased by 38 percent, the small ruminant inventory decreased 42 percent, the cattle-and-calves inventory decreased 17 percent, and methane emissions from livestock decreased 18 percent; yet despite the reduction in cattle numbers, US beef production increased over that period.

Some impacts of meat-producing livestock may be considered environmentally beneficial. These include waste reduction by conversion of human-inedible crop residues to food, use of livestock as an alternative to herbicides for control of invasive and noxious weeds and other vegetation management, use of animal manure as fertilizer as a substitute for those synthetic fertilizers that require considerable fossil fuel use for manufacture, grazing use for wildlife habitat enhancement, and carbon sequestration in response to grazing practices, among others.

Palm Oil

Palm oil, produced from the oil palm, is a basic source of income for many farmers in Southeast Asia, Central and West Africa, and Central America. It is locally used as a cooking oil, exported for use in many commercial food and personal care products and is converted into biofuel. It produces up to 10 times more oil per unit area as soyabeans, rapeseed or sunflowers. Oil palms produce 38% of vegetable oil output on 5% of the world's vegetable-oil farmland. Palm oil is under increasing scrutiny in relation to its effects on the environment.

A village palm oil press *"malaxeur"* in Bandundu, Democratic Republic of the Congo

Introductions and Invasive Species

Introductions of species, particularly plants into new areas, by whatever means and for whatever reasons have brought about major and permanent changes to the environment over large areas. Examples include the introduction of Caulerpa taxifolia into the Mediterranean, the introduction of oat species into the California grasslands, and the introduction of privet, kudzu, and purple loosestrife to North America. Rats, cats, and goats have radically altered biodiversity in many islands. Additionally, introductions have resulted in genetic changes to native fauna where interbreeding has taken place, as with buffalo with domestic cattle, and wolves with domestic dogs.

Energy Industry

The environmental impact of energy harvesting and consumption is diverse. In recent years there has been a trend towards the increased commercialization of various renewable energy sources.

In the real world, consumption of fossil fuel resources leads to global warming and climate change. However, little change is being made in many parts of the world. If the peak oil theory proves true, more explorations of viable alternative energy sources, could be more friendly to the environment.

Rapidly advancing technologies can achieve a transition of energy generation, water and waste management, and food production towards better environmental and energy usage practices using methods of systems ecology and industrial ecology.

Biodiesel

The environmental impact of biodiesel includes energy use, greenhouse gas emissions and some other kinds of pollution. A joint life cycle analysis by the US Department of Agriculture and the US Department of Energy found that substituting 100 percent biodiesel for petroleum diesel in buses reduced life cycle consumption of petroleum by 95 percent. Biodiesel reduced net emissions of carbon dioxide by 78.45 percent, compared with petroleum diesel. In urban buses, biodiesel reduced particulate emissions 32 percent, carbon monoxide emissions 35 percent, and emissions of sulfur oxides 8 percent, relative to life cycle emissions associated with use of petroleum diesel. Life cycle emissions of hydrocarbons were 35 percent higher and emission of various nitrogen oxides (NOx) were 13.5 percent higher with biodiesel. Life cycle analyses by the Argonne National Laboratory have indicated reduced fossil energy use and reduced greenhouse gas emissions with biodiesel, compared with petroleum diesel use. Biodiesel derived from various vegetable oils (e.g. canola or soybean oil), is readily biodegradable in the environment compared with petroleum diesel.

Coal Mining and Burning

The environmental impact of coal mining and -burning is diverse. Legislation passed by the US Congress in 1990 required the United States Environmental Protection Agency (EPA) to issue a plan to alleviate toxic air pollution from coal-fired power plants. After delay and litigation, the EPA now has a court-imposed deadline of March 16, 2011, to issue its report.

Electricity Generation

The environmental impact of electricity generation is significant because modern society uses large amounts of electrical power. This power is normally generated at power plants that convert some other kind of energy into electricity. Each such system has advantages and disadvantages, but many of them pose environmental concerns.

Nuclear Power

The environmental impact of nuclear power results from the nuclear fuel cycle processes including mining, processing, transporting and storing fuel and radioactive fuel waste. Released radioisotopes pose a health danger to human populations, animals and plants as radioactive particles enter organisms through various transmission routes.

Radiation is a carcinogen and causes numerous effects on living organisms and systems. The environmental impacts of nuclear power plant disasters such as the Chernobyl disaster, the Fukushima Daiichi nuclear disaster and the Three Mile Island accident, among others, persist indefinitely, though several other factors contributed to these events including improper management of fail safe systems and natural disasters putting uncommon stress on the generators. The radioactive decay rate of particles varies greatly, dependent upon the nuclear properties of a particular isotope. Radioactive Plutonium-244 has a half-life of 80.8 million years, which indicates the time duration required for half of a given sample to decay, though very little plutonium-244 is produced in the nuclear fuel cycle and lower half-life materials have lower activity thus giving off less dangerous radiation.

Oil Shale Industry

Kiviõli Oil Shale Processing & Chemicals Plant in ida-Virumaa, Estonia

The environmental impact of the oil shale industry includes the consideration of issues such as land use, waste management, and water and air pollution caused by the extraction and processing of oil shale. Surface mining of oil shale deposits causes the usual environmental impacts of open-pit mining. In addition, the combustion and thermal processing generate waste material, which must be disposed of, and harmful atmospheric emissions, including carbon dioxide, a major greenhouse gas. Experimental in-situ conversion processes and carbon capture and storage technologies may reduce some of these concerns in future, but may raise others, such as the pollution of groundwater.

Petroleum

The environmental impact of petroleum is often negative because it is toxic to almost all forms of life. Climate change exists. Petroleum, commonly referred to as oil, is closely linked to virtually all aspects of present society, especially for transportation and heating for both homes and for commercial activities.

Reservoirs

The Wachusett Dam in Clinton, Massachusetts.

The environmental impact of reservoirs is coming under ever increasing scrutiny as the world demand for water and energy increases and the number and size of reservoirs increases.

Dams and the reservoirs can be used to supply drinking water, generate hydroelectric power, increasing the water supply for irrigation, provide recreational opportunities and flood control. However, adverse environmental and sociological impacts have also been identified during and after many reservoir constructions. Although the impact varies greatly between different dams and reservoirs, common criticisms include preventing sea-run fish from reaching their historical mating grounds, less access to water downstream, and a smaller catch for fishing communities in the area. Advances in technology have provided solutions to many negative impacts of dams but these advances are often not viewed as worth investing in if not required by law or under the threat of fines. Whether reservoir projects are ultimately beneficial or detrimental—to both the environment and surrounding human populations— has been debated since the 1960s and probably long before that. In 1960 the construction of Llyn Celyn and the flooding of Capel Celyn provoked political uproar which continues to this day. More recently, the construction of Three Gorges Dam and other similar projects throughout Asia, Africa and Latin America have generated considerable environmental and political debate.

Wind Power

Compared to the environmental impact of traditional energy sources, the environmental impact of wind power is relatively minor. Wind powered electricity generation consumes no fuel, and emits no air pollution, unlike fossil fuel power sources. The energy consumed to manufacture and transport the materials used to build a wind power plant is equal to the new energy produced by the plant within a few months. While a wind farm may cover a large area of land, many land uses such as agriculture are compatible, with only small areas of turbine foundations and infrastructure made unavailable for use.

Wind turbines in an agricultural setting.

There are reports of bird and bat mortality at wind turbines, as there are around other artificial structures. The scale of the ecological impact may or may not be significant, depending on specific circumstances. Prevention and mitigation of wildlife fatalities, and protection of peat bogs, affect the siting and operation of wind turbines.

There are conflicting reports about the effects of noise on people who live very close to a wind turbine.

Light Pollution

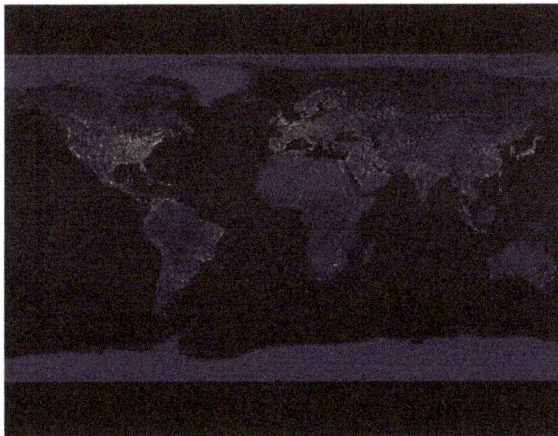

A composite image of artificial light emissions from Earth at night

Artificial light at night is one of the most obvious physical changes that humans have made to the biosphere, and is the easiest form of pollution to observe from space. The main environmental impacts of artificial light are due to light's use as an information source (rather than an energy source). The hunting efficiency of visual predators generally increases under artificial light, changing predator prey interactions. Artificial light also affects dispersal, orientation, migration, and hormone levels, resulting in disrupted circadian rhythms.

Manufactured Products

Cleaning Agents

The environmental impact of cleaning agents is diverse. In recent years, measures have been taken to reduce these effects.

Nanotechnology

Nanotechnology's environmental impact can be split into two aspects: the potential for nanotechnological innovations to help improve the environment, and the possibly novel type of pollution that nanotechnological materials might cause if released into the environment. As nanotechnology is an emerging field, there is great debate regarding to what extent industrial and commercial use of nanomaterials will affect organisms and ecosystems.

Leather

Paint

The environmental impact of paint is diverse. Traditional painting materials and processes can have harmful effects on the environment, including those from the use of lead and other additives. Measures can be taken to reduce environmental impact, including accurately estimating paint quantities so that wastage is minimized, use of paints, coatings, painting accessories and techniques that are environmentally preferred. The United States Environmental Protection Agency guidelines and Green Star ratings are some of the standards that can be applied.

Paper

The environmental impact of paper is significant, which has led to changes in industry and behaviour at both business and personal levels. With the use of modern technology such as the printing press and the highly mechanised harvesting of wood, paper has become a cheap commodity. This has led to a high level of consumption and waste. With the rise in environmental awareness due to the lobbying by environmental organizations and with increased government regulation there is now a trend towards sustainability in the pulp and paper industry.

A pulp and paper mill in New Brunswick, Canada. Although pulp and paper manufacturing requires large amounts of energy, a portion of it comes from burning wood waste.

Pesticides

The environmental impact of pesticides is often greater than what is intended by those who use them. Over 98% of sprayed insecticides and 95% of herbicides reach a destination other than their target species, including nontarget species, air, water, bottom sediments, and food. Pesticide contaminates land and water when it escapes from production sites and storage tanks, when it runs off from fields, when it is discarded, when it is sprayed aerially, and when it is sprayed into water to kill algae.

The amount of pesticide that migrates from the intended application area is influenced by the particular chemical's properties: its propensity for binding to soil, its vapor pressure, its water solubility, and its resistance to being broken down over time. Factors in the soil, such as its texture, its ability to retain water, and the amount of organic matter contained in it, also affect the amount of pesticide that will leave the area. Some pesticides contribute to global warming and the depletion of the ozone layer.

Pharmaceuticals and Personal Care Products

The environmental impact of pharmaceuticals and personal care products (PPCPs) is largely speculative. PPCPs are substances used by individuals for personal health or cosmetic reasons and the products used by agribusiness to boost growth or health of livestock. PPCPs have been detected in water bodies throughout the world. The effects of these chemicals on humans and the environment are not yet known, but to date there is no scientific evidence that they affect human health.

Mining

The environmental impact of mining includes erosion, formation of sinkholes, loss of biodiversity, and contamination of soil, groundwater and surface water by chemicals from mining processes. In some cases, additional forest logging is done in the vicinity of mines to increase the available room for the storage of the created debris and soil. Besides creating environmental damage, the contamination resulting from leakage of chemicals also affect the health of the local population. Mining companies in some countries are required to follow environmental and rehabilitation codes, ensuring the area mined is returned to close to its original state. Some mining methods may have significant environmental and public health effects.

Acid mine drainage in the Rio Tinto River.

Transport

The environmental impact of transport is significant because it is a major user of energy, and burns most of the world's petroleum. This creates air pollution, including nitrous oxides and particulates, and is a significant contributor to global warming through emission of carbon dioxide, for which transport is the fastest-growing emission sector. By subsector, road transport is the largest contributor to global warming.

Interstate 10 and Interstate 45 near downtown Houston, Texas in the United States.

Environmental regulations in developed countries have reduced the individual vehicles emission; however, this has been offset by an increase in the number of vehicles, and more use of each vehicle. Some pathways to reduce the carbon emissions of road vehicles considerably have been studied. Energy use and emissions vary largely between modes, causing environmentalists to call for a transition from air and road to rail and human-powered transport, and increase transport electrification and energy efficiency.

Other environmental impacts of transport systems include traffic congestion and automobile-oriented urban sprawl, which can consume natural habitat and agricultural lands. By reducing transportation emissions globally, it is predicted that there will be significant positive effects on Earth's air quality, acid rain, smog and climate change.

The health impact of transport emissions is also of concern. A recent survey of the studies on the effect of traffic emissions on pregnancy outcomes has linked exposure to emissions to adverse effects on gestational duration and possibly also intrauterine growth.

Aviation

The environmental impact of aviation occurs because aircraft engines emit noise, particulates, and gases which contribute to climate change and global dimming. Despite emission reductions from automobiles and more fuel-efficient and less polluting turbofan and turboprop engines, the rapid growth of air travel in recent years contributes to an increase in total pollution attributable to aviation. In the EU, greenhouse gas emissions from aviation increased by 87% between 1990 and 2006. Among other factors leading to this phenomenon are the increasing number of hypermobile travellers and social factors that are making air travel commonplace, such as frequent flyer programs.

There is an ongoing debate about possible taxation of air travel and the inclusion of aviation in an emissions trading scheme, with a view to ensuring that the total external costs of aviation are taken into account.

Roads

The environmental impact of roads includes the local effects of highways (public roads) such as on noise, light pollution, water pollution, habitat destruction/disturbance and local air quality; and the wider effects including climate change from vehicle emissions. The design, construction and management of roads, parking and other related facilities as well as the design and regulation of vehicles can change the impacts to varying degrees.

Shipping

The environmental impact of shipping includes greenhouse gas emissions and oil pollution. In 2007, carbon dioxide emissions from shipping were estimated at 4 to 5% of the global total, and estimated by the International Maritime Organisation (IMO) to rise by up to 72% by 2020 if no action is taken. There is also a potential for introducing invasive species into new areas through shipping, usually by attaching themselves to the ship's hull.

The First Intersessional Meeting of the IMO Working Group on Greenhouse Gas Emissions from Ships took place in Oslo, Norway on 23–27 June 2008. It was tasked with developing the technical basis for the reduction mechanisms that may form part of a future IMO regime to control greenhouse gas emissions from international shipping, and a draft of the actual reduction mechanisms themselves, for further consideration by IMO's Marine Environment Protection Committee (MEPC).

War

An Agent Orange spray run, part of Operation Ranch Hand, during the Vietnam War by UC-123B Provider aircraft.

As well as the cost to human life and society, there is a significant environmental impact of war. Scorched earth methods during, or after war have been in use for much of recorded history but with modern technology war can cause a far greater devastation on the environment. Unexploded ordnance can render land unusable for further use or make access across it dangerous or fatal.

Effects

Biodiversity

Human impact on biodiversity is significant, humans have caused the extinction of many species, including the dodo and, potentially, large megafaunal species during the last ice age. Though most experts agree that human beings have accelerated the rate of species extinction, the exact degree of this impact is unknown, perhaps 100 to 1000 times the normal background rate of extinction. Some authors have postulated that without human interference the biodiversity of the Earth would continue to grow at an exponential rate.

Coral Reefs

Human impact on coral reefs is significant. Coral reefs are dying around the world. In particular, coral mining, pollution (organic and non-organic), overfishing, blast fishing and the digging of canals and access into islands and bays are serious threats to these ecosystems. Coral reefs also face high dangers from pollution, diseases, destructive fishing practices and warming oceans. In order to find answers for these problems, researchers study the various factors that impact reefs. The list of factors is long, including the ocean's role as a carbon dioxide sink, atmospheric changes, ultraviolet light, ocean acidification, biological virus, impacts of dust storms carrying agents to far flung reefs, pollutants, algal blooms and others. Reefs are threatened well beyond coastal areas.

General estimates show approximately 10% world's coral reefs are already dead. It is estimated that about 60% of the world's reefs are at risk due to destructive, human-related activities. The threat to the health of reefs is particularly strong in Southeast Asia, where 80% of reefs are endangered.

Carbon Cycle

Global warming is the result of increasing atmospheric carbon dioxide concentrations which is caused primarily by the combustion of fossil energy sources such as petroleum, coal, and natural gas, and to an unknown extent by destruction of forests, increased methane, volcanic activity and cement production. Such massive alteration of the global carbon cycle has only been possible because of the availability and deployment of advanced technologies, ranging in application from fossil fuel exploration, extraction, distribution, refining, and combustion in power plants and automobile engines and advanced farming practices. Livestock contributes to climate change both thru the production of greenhouse gases and thru destruction of carbon sinks such as rain-forests. According to the 2006 United Nations/FAO report, 18% of all greenhouse gas emissions found in the atmosphere are due to livestock. The raising of livestock and the land needed to feed them has resulted in the destruction millions of acres of Rainforest and as global demand for meat rises, so too will the demand for land. Ninety-one percent of all rainforest land deforested since 1970 is now used for livestock. Potential negative environmental impacts caused by increasing atmospheric carbon dioxide concentrations are rising global air temperatures, altered hydrogeological cycles resulting in more frequent and severe droughts, storms, and floods, as well as sea level rise and ecosystem disruption.

Nitrogen Cycle

Human impact on the nitrogen cycle is diverse. Agricultural and industrial nitrogen (N) inputs to the environment currently exceed inputs from natural N fixation. As a consequence of anthropogenic inputs, the global nitrogen cycle (Fig. 1) has been significantly altered over the past century. Global atmospheric nitrous oxide (N_2O) mole fractions have increased from a pre-industrial value of ~270 nmol/mol to ~319 nmol/mol in 2005. Human activities account for over one-third of N_2O emissions, most of which are due to the agricultural sector.

Human Impact on the Nitrogen Cycle

Human impact on the nitrogen cycle is diverse. Agricultural and industrial nitrogen (N) inputs to the environment currently exceed inputs from natural N fixation. As a consequence of anthropogenic inputs, the global nitrogen cycle (Fig. 1) has been significantly altered over the past century. Global atmospheric nitrous oxide (N_2O) mole fractions have increased from a pre-industrial value of ~270 nmol/mol to ~319 nmol/mol in 2005. Human activities account for over one-third of N_2O emissions, most of which are due to the agricultural sector. This article is intended to give a brief review of the history of anthropogenic N inputs, and reported impacts of nitrogen inputs on selected terrestrial and aquatic ecosystems.

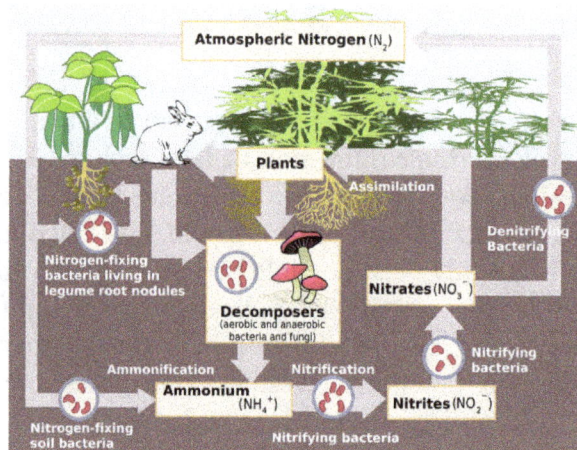

Figure 1. The nitrogen cycle in a soil-plant system. One potential pathway: N is fixed by microbes into organic compounds, which are mineralized (i.e., ammonification) and then oxidized to inorganic forms (i.e., nitrification) that are assimilated by plants (NO_3^-). NO_3^- may also be denitrified by bacteria, producing N_2, NO_x, and N_2O.

History of Anthropogenic Nitrogen Inputs

Approximately 78% of earth's atmosphere is N gas (N_2), which is an inert compound and biologically unavailable to most organisms. In order to be utilized in most biological processes, N_2 must be converted to reactive N (Nr), which includes inorganic reduced forms (NH_3 and NH_4^+), inorganic oxidized forms (NO, NO_2, HNO_3, N_2O, and NO_3^-), and organic compounds (urea, amines, and proteins). N_2 has a strong triple bond, and so a significant amount of energy (226 kcal mol-1) is required to convert N_2 to Nr. Prior to industrial processes, the only sources of such energy were solar radiation and electrical discharges. Utilizing a large amount of metabolic energy and the enzyme nitrogenase, some bacteria and cyanobacteria convert atmospheric N_2 to NH_3, a process known as biological nitrogen fixation (BNF). The anthropogenic analogue to BNF is the Haber-Bosch process, in which fossil fuel H_2 is reacted with atmospheric N_2 at high temperatures and pressures to produce NH_3. Lastly, N_2 is converted to NO by energy from lightning, which is negligible in current temperate ecosystems, or by fossil fuel combustion.

Until 1850, natural BNF, cultivation-induced BNF (e.g., planting of leguminous crops), and incorporated organic matter were the only sources of N for agricultural production. Near the turn of the century, Nr from guano and sodium nitrate deposits was harvested and exported from the arid Pacific islands and South American deserts. By the late 1920s, early industrial processes, albeit inefficient, were commonly used to produce NH_3. Due to the efforts of Fritz Haber and Carl Bosch, the Haber-Bosch process became the largest source of nitrogenous fertilizer after the 1950s, and replaced BNF as the dominant source of NH_3 production. From 1890 to 1990, anthropogenically created Nr increased almost ninefold. During this time, mango population more than tripled, partly due to increased food production.

Since the industrial revolution, an additional source of anthropogenic N input has been fossil fuel combustion, which is used to generate energy (e.g., to power automobiles). During combustion of fossil fuels, high temperatures and pressures provide energy to produce NO from N_2 oxidation. Additionally, when fossil fuel is extracted and burned, fossil N may become reactive (i.e., NO_x emissions). During the 1970s, scientists began to recognize that N inputs were accumulating in the environment and affecting ecosystem functioning.

Impacts of Anthropogenic Inputs on the Nitrogen Cycle

Between 1600 and 1990, global reactive nitrogen (Nr) creation had increased nearly 50%. During this period, atmospheric emissions of Nr species reportedly increased 250% and deposition to marine and terrestrial ecosystems increased over 200%. Additionally, there was a reported fourfold increase in riverine dissolved inorganic N fluxes to coasts. Nitrogen is a critical limiting nutrient in many systems, including forests, wetlands, and coastal and marine ecosystems; therefore, this change in emissions and distribution of Nr has resulted in substantial consequences for aquatic and terrestrial ecosystems.

Atmosphere

Atmospheric N inputs mainly include oxides of N (NO_x), ammonia (NH_3), and nitrous oxide (N_2O) from aquatic and terrestrial ecosystems, and NO_x from fossil fuel and biomass combustion.

In agroecosystems, fertilizer application has increased microbial nitrification (aerobic process in which microorganisms oxidize ammonium [NH_4^+] to nitrate [NO_3^-]) and denitrification (anaerobic process in which microorganisms reduce NO_3^- to atmospheric nitrogen gas [N_2]). Both processes naturally leak nitric oxide (NO) and nitrous oxide (N_2O) to the atmosphere. Of particular concern is N_2O, which has an average atmospheric lifetime of 114–120 years, and is 300 times more effective than CO_2 as a greenhouse gas. NO_x produced by industrial processes, automobiles and agricultural fertilization and NH_3 emitted from soils (i.e., as an additional byproduct of nitrification) and livestock operations are transported to downwind ecosystems, influencing N cycling and nutrient losses. Six major effects of NO_x and NH_3 emissions have been cited: 1) decreased atmospheric visibility due to ammonium aerosols (fine particulate matter [PM]); 2) elevated ozone concentrations; 3) ozone and PM affects human health (e.g. respiratory diseases, cancer); 4) increases in radiative forcing and global climate change; 5) decreased agricultural productivity due to ozone deposition; and 6) ecosystem acidification and eutrophication.

Biosphere

Terrestrial and aquatic ecosystems receive Nr inputs from the atmosphere through wet and dry deposition. Atmospheric Nr species can be deposited to ecosystems in precipitation (e.g., NO_3^-, NH_4^+, organic N compounds), as gases (e.g., NH_3 and gaseous nitric acid [HNO_3]), or as aerosols (e.g., ammonium nitrate [NH_4NO_3]). Aquatic ecosystems receive additional nitrogen from surface runoff and riverine inputs.

Increased N deposition can acidify soils, streams, and lakes and alter forest and grassland productivity. In grassland ecosystems, N inputs have produced initial increases in productivity followed by declines as critical thresholds are exceeded. Nitrogen effects on biodiversity, carbon cycling, and changes in species composition have also been demonstrated. In highly developed areas of near shore coastal ocean and estuarine systems, rivers deliver direct (e.g., surface runoff) and indirect (e.g., groundwater contamination) N inputs from agroecosystems. Increased N inputs can result in freshwater acidification and eutrophication of marine waters.

Terrestrial Ecosystems

Impacts on Productivity and Nutrient Cycling

Much of terrestrial growth in temperate systems is limited by N; therefore, N inputs (i.e., through deposition and fertilization) can increase N availability, which temporarily increases N uptake, plant and microbial growth, and N accumulation in plant biomass and soil organic matter. Incorporation of greater amounts of N in organic matter decreases C:N ratios, increasing mineral N release (NH_4^+) during organic matter decomposition by heterotrophic microbes (i.e.ammonification). As ammonification increases, so does nitrification of the mineralized N. Because microbial nitrification and denitrification are "leaky", N deposition is expected to increase trace gas emissions. Additionally, with increasing NH_4^+ accumulation in the soil, nitrification processes release hydrogen ions, which acidify the soil. NO_3^-, the product of nitrification, is highly mobile and can be leached from the soil, along with positively charged alkaline minerals such as calcium and magnesium. In acid soils, mobilized aluminium ions can reach toxic concentrations, negatively affecting both terrestrial and adjacent aquatic ecosystems.

Anthropogenic sources of N generally reach upland forests through deposition. A potential concern of increased N deposition due to human activities is altered nutrient cycling in forest ecosystems. Numerous studies have demonstrated both positive and negative impacts of atmospheric N deposition on forest productivity and carbon storage. Added N is often rapidly immobilized by microbes, and the effect of the remaining available N depends on the plant community's capacity for N uptake. In systems with high uptake, N is assimilated into the plant biomass, leading to enhanced net primary productivity (NPP) and possibly increased carbon sequestration through greater photosynthetic capacity. However, ecosystem responses to N additions are contingent upon many site-specific factors including climate, land-use history, and amount of N additions. For example, in the Northeastern United States, hardwood stands receiving chronic N inputs have demonstrated greater capacity to retain N and increase annual net primary productivity (ANPP) than conifer stands. Once N input exceeds system demand, N may be lost via leaching and gas fluxes. When available N exceeds the ecosystem's (i.e., vegetation, soil, and microbes, etc.) uptake capacity, N saturation occurs and excess N is lost to surface waters, groundwater, and the atmosphere. N saturation can result in nutrient imbalances (e.g., loss of calcium due to nitrate leaching) and possible forest decline.

A 15-year study of chronic N additions at the Harvard Forest Long Term Ecological Research (LTER) program has elucidated many impacts of increased nitrogen deposition on nutrient cycling in temperate forests. It found that chronic N additions resulted in greater leaching losses, increased pine mortality, and cessation of biomass accumulation. Another study reported that chronic N additions resulted in accumulation of non-photosynthetic N and subsequently reduced photosynthetic capacity, supposedly leading to severe carbon stress and mortality. These findings negate previous hypotheses that increased N inputs would increase NPP and carbon sequestration.

Impacts on Plant Species Diversity

Many plant communities have evolved under low nutrient conditions; therefore, increased N inputs can alter biotic and abiotic interactions, leading to changes in community composition. Several nutrient addition studies have shown that increased N inputs lead to dominance of fast-growing plant species, with associated declines in species richness. Other studies have found that secondary responses of the system to N enrichment, including soil acidification and changes in mycorrhizal communities have allowed stress-tolerant species to out-compete sensitive species. Two other studies found evidence that increased N availability has resulted in declines in species-diverse heathlands. Heathlands are characterized by N-poor soils, which exclude N-demanding grasses; however, with increasing N deposition and soil acidification, invading grasslands replace lowland heath.

In a more recent experimental study of N fertilization and disturbance (i.e., tillage) in old field succession, it was found that species richness decreased with increasing N, regardless of disturbance level. Competition experiments showed that competitive dominants excluded competitively inferior species between disturbance events. With increased N inputs, competition shifted from belowground to aboveground (i.e., to competition for light), and patch colonization rates significantly decreased. These internal changes can dramatically affect the community by shifting the balance of competition-colonization tradeoffs between species. In patch-based systems, regional coexistence can occur through tradeoffs in competitive and colonizing abilities given sufficiently high disturbance rates. That is, with inverse ranking of competitive and colonizing abilities, plants can coexist in space and time as disturbance removes superior competitors from patches, allowing for establishment of superior colonizers. However, as demonstrated by Wilson and Tilman, increased nutrient inputs can negate tradeoffs, resulting in competitive exclusion of these superior colonizers/poor competitors.

Aquatic Ecosystems

Aquatic ecosystems also exhibit varied responses to nitrogen enrichment. NO_3^- loading from N saturated, terrestrial ecosystems can lead to acidification of downstream freshwater systems and eutrophication of downstream marine systems. Freshwater acidification can cause aluminium toxicity and mortality of pH-sensitive fish species. Because marine systems are generally nitrogen-limited, excessive N inputs can result in water quality degradation due to toxic algal blooms, oxygen deficiency, habitat loss, decreases in biodiversity, and fishery losses.

Acidification of Freshwaters

Atmospheric N deposition in terrestrial landscapes can be transformed through soil microbial processes to biologically available nitrogen, which can result in surface-water acidification, and loss of biodiversity. NO_3^- and NH_4^+ inputs from terrestrial systems and the atmosphere can acidify freshwater systems when there is little buffering capacity due to soil acidification. N pollution in Europe, the Northeastern United States, and Asia is a current concern for freshwater acidification. Lake acidification studies in the Experimental Lake Area (ELA) in northwestern Ontario clearly demonstrated the negative effects of increased acidity on a native fish species: lake trout (Salvelinus namaycush) recruitment and growth dramatically decreased due to extirpation of its key prey species during acidification.

Eutrophication of Marine Systems

Urbanization, deforestation, and agricultural activities largely contribute sediment and nutrient inputs to coastal waters via rivers. Increased nutrient inputs to marine systems have shown both short-term increases in productivity and fishery yields, and long-term detrimental effects of eutrophication. Tripling of NO_3^- loads in the Mississippi River in the last half of the 20th century have been correlated with increased fishery yields in waters surrounding the Mississippi delta; however, these nutrient inputs have produced seasonal hypoxia (oxygen concentrations less than 2–3 mg L^{-1}, "dead zones") in the Gulf of Mexico. In estuarine and coastal systems, high nutrient inputs increase primary production (e.g., phytoplankton, sea grasses, macroalgae), which increase turbidity with resulting decreases in light penetration throughout the water column. Consequently, submerged vegetation growth declines, which reduces habitat complexity and oxygen production. The increased primary (i.e., phytoplankton, macroalgae, etc.) production leads to a flux of carbon to bottom waters when decaying organic matter (i.e., senescent primary production) sinks and is consumed by aerobic bacteria lower in the water column. As a result, oxygen consumption in bottom waters is greater than diffusion of oxygen from surface waters .

Integration

The above system responses to reactive nitrogen (Nr) inputs are almost all exclusively studied separately; however, research increasingly indicates that nitrogen loading problems are linked by multiple pathways transporting nutrients across system boundaries. This sequential transfer between ecosystems is termed the nitrogen cascade. During the cascade, some systems accumulate Nr, which results in a time lag in the cascade and enhanced effects of Nr on the environment in which it accumulates. Ultimately, anthropogenic inputs of Nr are either accumulated or denitrified; however, little progress has been made in determining the relative importance of Nr accumulation and denitrification, which has been mainly due to a lack of integration among scientific disciplines.

Most Nr applied to global agroecosystems cascades through the atmosphere and aquatic and terrestrial ecosystems until it is converted to N_2, primarily through denitrification. Although terrestrial denitrification produces gaseous intermediates (nitric oxide [NO] and nitrous oxide [N_2O]), the last step—microbial production of N_2—is critical because atmospheric N_2 is a sink for Nr. Many studies have clearly demonstrated that managed buffer strips and wetlands can remove significant amounts of nitrate (NO_3^-) from agricultural systems through denitrification. Such management may help attenuate the undesirable cascading effects and eliminate environmental Nr accumulation.

Human activities dominate the global and most regional N cycles. N inputs have shown negative consequences for both nutrient cycling and native species diversity in terrestrial and aquatic systems. In fact, due to long-term impacts on food webs, Nr inputs are widely considered the most critical pollution problem in marine systems. In both terrestrial and aquatic ecosystems, responses to N enrichment vary; however, a general re-occurring theme is the importance of thresholds (e.g., nitrogen saturation) in system nutrient retention capacity. In order to control the N cascade, there must be integration of scientific disciplines and further work on Nr storage and denitrification rates.

I = PAT

I = PAT is the lettering of a formula put forward to describe the impact of human activity on the environment.

$$I = P \times A \times T$$

In words:

> Human Impact on the environment equals the product of Population, Affluence, and Technology. This shows how the population, affluence and technology produce an impact

The equation was developed in the 1970s during the course of a debate between Barry Commoner, Paul R. Ehrlich and John Holdren. Commoner argued that environmental impacts in the United States were caused primarily by changes in its production technology following World War II, while Ehrlich and Holdren argued that all three factors were important and emphasized in particular the role of human population growth.

The equation can aid in understanding some of the factors affecting human impacts on the environment, but it has also been cited as one of the primary factors underlying many of the dire environmental predictions of the 1970s by Paul Ehrlich, George Wald, Denis Hayes, Lester Brown, René Dubos, and Sidney Ripley that did not come to pass. Neal Koblitz classified equations of this type as "mathematical propaganda" and criticized Ehrlich's use of them in the media (e.g. on The Tonight Show) to sway the general public.

The Kaya identity is closely related to the I = PAT equation. The I = PAT equation is more general, describing an abstract "impact". The Kaya identity describes more clearly the impact of human activity on CO_2 emissions.

Population

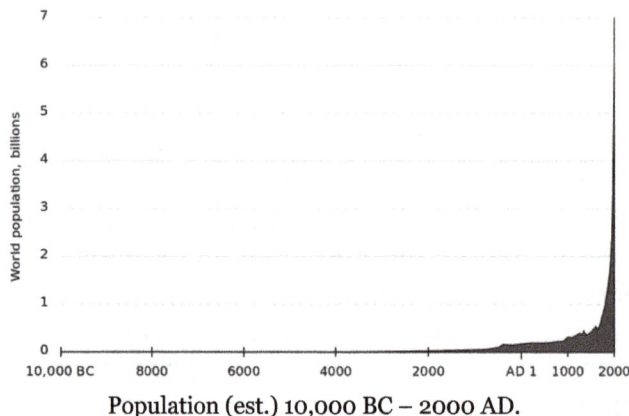

Population (est.) 10,000 BC – 2000 AD.

In the I=PAT equation, the variable P represents the population of an area, such as the world. Since the rise of industrial societies, human population has been increasing exponentially. This has caused Thomas Malthus and many others to postulate that this growth would continue until checked by widespread hunger and famine.

The United Nations and the US Census Bureau project that world population will increase from 7.4 billion today in 2016 (up from 7.0 billion forecast in 2005) to about 9.7 billion by 2050 (up from 9.2 billion). These projections take into consideration that population growth has slowed in recent years as women are having fewer children. This phenomenon is the result of demographic transition all over the world. The UN projects that human population might stabilize around 11.2 billion by 2100 (up from 9.2 billion). However, since the world population is set to keep rising for the next few decades, this factor of the I=PAT equation will likely keep increasing human impact on the environment for the near future.

Environmental Impacts

Increased population increases humans' environmental impact in many ways, which include but are not limited to:

- Increased land use - Results in habitat loss for other species.

- Increased resource use - Results in changes in land cover

- Increased pollution - Can cause sickness and damages ecosystems.

Affluence

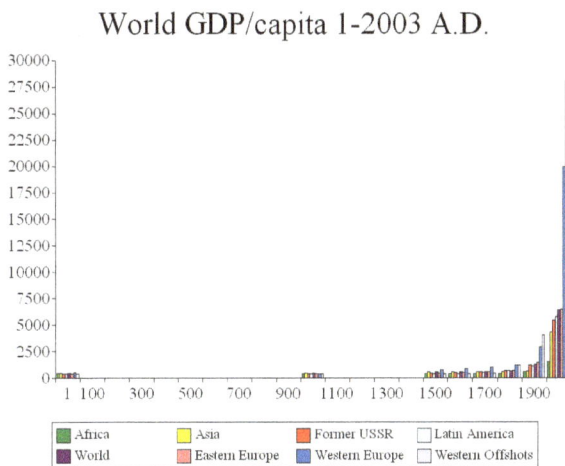

World GDP per capita (in 1990 Geary-Khamis dollars)

The variable A, in the I=PAT equation stands for affluence. It represents the average consumption of each person in the population. As the consumption of each person increases, the total environmental impact increases as well. A common proxy for measuring consumption is through GDP per capita. While GDP per capita measures production, it is often assumed that consumption increases when production increases. GDP per capita has been rising steadily over the last few centuries and is driving up human impact in the I=PAT equation.

Environmental Impacts

Increased consumption significantly increases human environmental impact. This is because each product consumed has wide ranging effects on the environment. For example, if the construction of a car had the following environmental impacts among others:

- 605,664 gallons of water for parts and tires

- 682 lbs. of pollution at a mine for the lead battery.

- 2178 lbs. of discharge into water supply for the 22 lbs. of copper contained in the car.

then the more cars per capita, the greater the impact. Since the ecological impacts of each product are far reaching, increases in consumption quickly result in large impacts on the environment.

Technology

The T variable in the I=PAT equation represents how resource intensive the production of affluence is; how much environmental impact is involved in creating, transporting and disposing of the goods, services and amenities used. Improvements in efficiency can reduce resource intensiveness, reducing the T multiplier. Since technology can affect environmental impact in many different ways, the unit for T is often tailored for the situation I=PAT is being applied to. For example, for a situation where the human impact on climate change is being measured, an appropriate unit for T might be greenhouse gas emissions per unit of GDP.

Environmental Impacts

Increases in efficiency can reduce overall environmental impact. However, since P has increased exponentially, and A has also increased drastically, the overall environmental impact, I, has still increased.

Reception

The I=PAT equation has been criticized for being too simplistic by assuming that P, A, and T are independent of each other. In reality, at least 7 interdependencies between P, A, and T could exist, indicating that it is more correct to rewrite the equation as I = f(P,A,T). For example, a doubling of technological efficiency, or equivalently a reduction of the T-factor by 50%, does not necessarily reduce the environmental impact (I) by 50% if efficiency induced price reductions stimulate additional consumption of the resource that was supposed to be conserved, a phenomenon called the rebound effect (conservation) or Jevons Paradox. As was shown by Alcott, despite significant improvements in the carbon intensity of GDP (i.e., the efficiency in carbon use) since 1980, world fossil energy consumption has increased in line with economic and population growth. Similarly, an extensive historical analysis of technological efficiency improvements has conclusively shown that energy and materials use efficiency improvements were almost always outpaced by economic growth, resulting in a net increase in resource use and associated pollution.

As a result of the interdependencies between P, A, and T and potential rebound effects, policies aimed at decreasing environmental impacts through reductions in P, A, and T may not only be very difficult to implement (i.e., population control and material sufficiency and degrowth movements have been very controversial) but also are likely to be rather ineffective compared to rationing (i.e., quotas) or Pigouvian taxation of resource use or pollution.

Life-cycle Assessment

the life cycle assessment

" lca is a consulting-based model that can be used to evaluate the resource usage and environmental effects of all the stages of a product, process or activity, to aid environmental decision-making. "

the general life cycle

Materials acquisition
Materials processing
Manufacturing
Assembly
Packaging
Transportation/ distribution
Product use
Reuse, Recycle, Disposal

lca methodology

Life Cycle Assessment Overview

Life-cycle assessment (LCA, also known as life-cycle analysis, ecobalance, and cradle-to-grave analysis) is a technique to assess environmental impacts associated with all the stages of a product's life from cradle to grave (i.e., from raw material extraction through materials processing, manufacture, distribution, use, repair and maintenance, and disposal or recycling). Designers use this process to help critique their products. LCAs can help avoid a narrow outlook on environmental concerns by:

- Compiling an inventory of relevant energy and material inputs and environmental releases;

- Evaluating the potential impacts associated with identified inputs and releases;

- Interpreting the results to help make a more informed decision.

Goals and Purpose

The goal of LCA is to compare the full range of environmental effects assignable to products and services by quantifying all inputs and outputs of material flows and assessing how these material flows affect the environment. This information is used to improve processes, support policy and provide a sound basis for informed decisions.

Life Cycle Assessment A systematic set of procedures for compiling and examining the inputs and outputs of materials and energy and the associated environmental impacts directly attributable to the functioning of a product or service system throughout its life cycle.

The term *life cycle* refers to the notion that a fair, holistic assessment requires the assessment of raw-material production, manufacture, distribution, use and disposal including all intervening transportation steps necessary or caused by the product's existence.

There are two main types of LCA. Attributional LCAs seek to establish (or attribute) the burdens associated with the production and use of a product, or with a specific service or process, at a point in time (typically the recent past). Consequential LCAs seek to identify the environmental consequences of a decision or a proposed change in a system under study (oriented to the future), which means that market and economic implications of a decision may have to be taken into account. Social LCA is under development as a different approach to life cycle thinking intended to assess social implications or potential impacts. Social LCA should be considered as an approach that is complementary to environmental LCA.

The procedures of life cycle assessment (LCA) are part of the ISO 14000 environmental management standards: in ISO 14040:2006 and 14044:2006. (ISO 14044 replaced earlier versions of ISO 14041 to ISO 14043.) GHG product life cycle assessments can also comply with specifications such as PAS 2050 and the GHG Protocol Life Cycle Accounting and Reporting Standard.

Four Main Phases

According to the ISO 14040 and 14044 standards, a Life Cycle Assessment is carried out in four distinct phases as illustrated in the figure shown to the right. The phases are often interdependent in that the results of one phase will inform how other phases are completed.

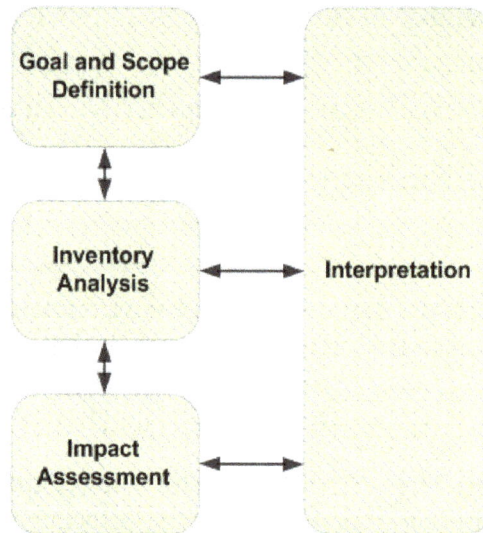

Illustration of LCA phases

Goal and Scope

An LCA starts with an explicit statement of the goal and scope of the study, which sets out the context of the study and explains how and to whom the results are to be communicated. This is a key step and the ISO standards require that the goal and scope of an LCA be clearly defined and consistent with the intended application. The goal and scope document therefore includes technical details that guide subsequent work:

- the functional unit, which defines what precisely is being studied and quantifies the service delivered by the product system, providing a reference to which the inputs and outputs can be related. Further, the functional unit is an important basis that enables alternative goods, or services, to be compared and analyzed. So to explain this a functional system which is inputs, processes and outputs contains a functional unit, that fulfills a function, for example paint is covering a wall, making a functional unit of 1m² covered for 10 years. The functional flow would be the items necessary for that function, so this would be a brush, tin of paint and the paint itself.

- the system boundaries;

- any assumptions and limitations;

- the allocation methods used to partition the environmental load of a process when several products or functions share the same process; allocation is commonly dealt with in one of three ways: system expansion, substitution and partition. Doing this is not easy and different methods may give different results and

- the impact categories chosen for example human toxicity, smog, global warming, eutrophication.

Life Cycle Inventory

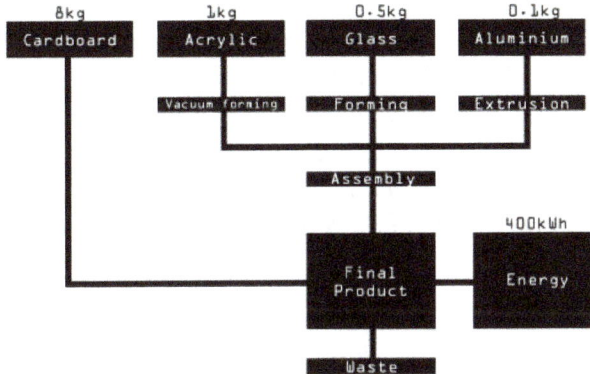

This is an example of a Life-cycle inventory (LCI) diagram

Life Cycle Inventory (LCI) analysis involves creating an inventory of flows from and to nature for a product system. Inventory flows include inputs of water, energy, and raw materials, and releases to air, land, and water. To develop the inventory, a flow model of the technical system is constructed using data on inputs and outputs. The flow model is typically illustrated with a flow chart that includes the activities that are going to be assessed in the relevant supply chain and gives a clear picture of the technical system boundaries. The input and output data needed for the construction of the model are collected for all activities within the system boundary, including from the supply chain (referred to as inputs from the techno-sphere).

The data must be related to the functional unit defined in the goal and scope definition. Data can be presented in tables and some interpretations can be made already at this stage. The results of the inventory is an LCI which provides information about all inputs and outputs in the form of elementary flow to and from the environment from all the unit processes involved in the study.

Inventory flows can number in the hundreds depending on the system boundary. For product LCAs at either the generic (i.e., representative industry averages) or brand-specific level, that data is typically collected through survey questionnaires. At an industry level, care has to be taken to ensure that questionnaires are completed by a representative sample of producers, leaning toward neither the best nor the worst, and fully representing any regional differences due to energy use, material sourcing or other factors. The questionnaires cover the full range of inputs and outputs, typically aiming to account for 99% of the mass of a product, 99% of the energy used in its production and any environmentally sensitive flows, even if they fall within the 1% level of inputs.

One area where data access is likely to be difficult is flows from the techno-sphere. The technosphere is more simply defined as the man-made world. Considered by geologists as secondary resources, these resources are in theory 100% recyclable; however, in a practical sense the primary goal is salvage. For an LCI, these technosphere products (supply chain products) are those that have been produced by man and unfortunately those completing a questionnaire about a process which uses man-made product as a means to an end will be unable to specify how much of a given input they use. Typically, they will not have access to data concerning inputs and outputs for previous production processes of the product. The entity undertaking the LCA must then turn to secondary sources if it does not already have that data from its own previous studies. National databases or data sets that come with LCA-practitioner tools, or that can be readily accessed, are the usual sources for that information. Care must then be taken to ensure that the secondary data source properly reflects regional or national conditions.

Life Cycle Impact Assessment

Inventory analysis is followed by impact assessment. This phase of LCA is aimed at evaluating the significance of potential environmental impacts based on the LCI flow results. Classical life cycle impact assessment (LCIA) consists of the following mandatory elements:

- selection of impact categories, category indicators, and characterization models;

- the classification stage, where the inventory parameters are sorted and assigned to specific impact categories; and

- impact measurement, where the categorized LCI flows are characterized, using one of many possible LCIA methodologies, into common equivalence units that are then summed to provide an overall impact category total.

In many LCAs, characterization concludes the LCIA analysis; this is also the last compulsory stage according to ISO 14044:2006. However, in addition to the above mandatory LCIA steps, other optional LCIA elements – normalization, grouping, and weighting – may be conducted depending on the goal and scope of the LCA study. In normalization, the results of the impact categories from the study are usually compared with the total impacts in the region of interest, the U.S. for example. Grouping consists of sorting and possibly ranking the impact categories. During weighting, the different environmental impacts are weighted relative to each other so that they can then be summed to get a single number for the total environmental impact. ISO 14044:2006 generally advises against weighting, stating that "weighting, shall not be used in LCA studies intended to be used in comparative assertions intended to be disclosed to the public". This advice is often ignored, resulting in comparisons that can reflect a high degree of subjectivity as a result of weighting.

Interpretation

Life Cycle Interpretation is a systematic technique to identify, quantify, check, and evaluate information from the results of the life cycle inventory and/or the life cycle impact assessment. The results from the inventory analysis and impact assessment are summarized during the interpretation phase. The outcome of the interpretation phase is a set of conclusions and recommendations for the study. According to ISO 14040:2006, the interpretation should include:

- identification of significant issues based on the results of the LCI and LCIA phases of an LCA;

- evaluation of the study considering completeness, sensitivity and consistency checks; and

- conclusions, limitations and recommendations.

A key purpose of performing life cycle interpretation is to determine the level of confidence in the final results and communicate them in a fair, complete, and accurate manner. Interpreting the results of an LCA is not as simple as "3 is better than 2, therefore Alternative A is the best choice"! Interpreting the results of an LCA starts with understanding the accuracy of the results, and ensuring they meet the goal of the study. This is accomplished by identifying the data elements that contribute significantly to each impact category, evaluating the sensitivity of these significant data elements, assessing the completeness and consistency of the study, and drawing conclusions and recommendations based on a clear understanding of how the LCA was conducted and the results were developed.

Reference Test

More specifically, the best alternative is the one that the LCA shows to have the least cradle-to-grave environmental negative impact on land, sea, and air resources.

LCA Uses

Based on a survey of LCA practitioners carried out in 2006 LCA is mostly used to support business strategy (18%) and R&D (18%), as input to product or process design (15%), in education (13%) and for labeling or product declarations (11%). LCA will be continuously integrated into the built environment as tools such as the European EN-SLIC Building project guidelines for buildings or developed and implemented, which provide practitioners guidance on methods to implement LCI data into the planning and design process.

Major corporations all over the world are either undertaking LCA in house or commissioning studies, while governments support the development of national databases to support LCA. Of particular note is the growing use of LCA for ISO Type III labels called Environmental Product Declarations, defined as "quantified environmental data for a product with pre-set categories of parameters based on the ISO 14040 series of standards, but not excluding additional environmental information". These third-party certified LCA-based labels provide an increasingly important basis for assessing the relative environmental merits of competing products. Third-party certification plays a major role in today's industry. Independent certification can show a company's dedication to safer and environmental friendlier products to customers and NGOs.

LCA also has major roles in environmental impact assessment, integrated waste management and pollution studies.

Data Analysis

A life cycle analysis is only as valid as its data; therefore, it is crucial that data used for the completion of a life cycle analysis are accurate and current. When comparing different life cycle analyses with one another, it is crucial that equivalent data are available for both products or processes in question. If one product has a much higher availability of data, it cannot be justly compared to another product which has less detailed data.

There are two basic types of LCA data – unit process data and environmental input-output data (EIO), where the latter is based on national economic input-output data. Unit process data are derived from direct surveys of companies or plants producing the product of interest, carried out at a unit process level defined by the system boundaries for the study.

Data validity is an ongoing concern for life cycle analyses. Due to globalization and the rapid pace of research and development, new materials and manufacturing methods are continually being introduced to the market. This makes it both very important and very difficult to use up-to-date information when performing an LCA. If an LCA's conclusions are to be valid, the data must be recent; however, the data-gathering process takes time. If a product and its related processes have not undergone significant revisions since the last LCA data was collected, data validity is not a problem. However, consumer electronics such as cell phones can be redesigned as often as every 9 to 12 months, creating a need for ongoing data collection.

The life cycle considered usually consists of a number of stages including: materials extraction, processing and manufacturing, product use, and product disposal. If the most environmentally harmful of these stages can be determined, then impact on the environment can be efficiently reduced by focusing on making changes for that particular phase. For example, the most energy-intensive life phase of an airplane or car is during use due to fuel consumption. One of the most effective ways to increase fuel efficiency is to decrease vehicle weight, and thus, car and airplane manufacturers can decrease environmental impact in a significant way by replacing heavier materials with lighter ones such as aluminium or carbon fiber-reinforced elements. The reduction during the use phase should be more than enough to balance additional raw material or manufacturing cost.

Variants

Cradle-to-grave

Cradle-to-grave is the full Life Cycle Assessment from resource extraction ('cradle') to use phase and disposal phase ('grave'). For example, trees produce paper, which can be recycled into low-energy production cellulose (fiberised paper) insulation, then used as an energy-saving device in the ceiling of a home for 40 years, saving 2,000 times the fossil-fuel energy used in its production. After 40 years the cellulose fibers are replaced and the old fibers are disposed of, possibly incinerated. All inputs and outputs are considered for all the phases of the life cycle.

Cradle-to-gate

Cradle-to-gate is an assessment of a *partial* product life cycle from resource extraction (*cradle*) to the factory gate (i.e., before it is transported to the consumer). The use phase and disposal phase of the product are omitted in this case. Cradle-to-gate assessments are sometimes the basis for environmental product declarations (EPD) termed business-to-business EDPs. One of the significant uses of the cradle-to-gate approach compiles the life cycle inventory (LCI) using cradle-to-gate. This allows the LCA to collect all of the impacts leading up to resources being purchased by the facility. They can then add the steps involved in their transport to plant and manufacture process to more easily produce their own cradle-to-gate values for their products.

Cradle-to-cradle or Closed Loop Production

Cradle-to-cradle is a specific kind of cradle-to-grave assessment, where the end-of-life disposal step for the product is a recycling process. It is a method used to minimize the environmental impact of products by employing sustainable production, operation, and disposal practices and aims to incorporate social responsibility into product development. From the recycling process originate new, identical products (e.g., asphalt pavement from discarded asphalt pavement, glass bottles from collected glass bottles), or different products (e.g., glass wool insulation from collected glass bottles).

Allocation of burden for products in open loop production systems presents considerable challenges for LCA. Various methods, such as the avoided burden approach have been proposed to deal with the issues involved.

Gate-to-gate

Gate-to-gate is a partial LCA looking at only one value-added process in the entire production chain. Gate-to-gate modules may also later be linked in their appropriate production chain to form a complete cradle-to-gate evaluation.

Well-to-wheel

Well-to-wheel is the specific LCA used for transport fuels and vehicles. The analysis is often broken down into stages entitled "well-to-station", or "well-to-tank", and "station-to-wheel" or "tank-to-wheel", or "plug-to-wheel". The first stage, which incorporates the feedstock or fuel production and processing and fuel delivery or energy transmission, and is called the "upstream" stage, while the stage that deals with vehicle operation itself is sometimes called the "downstream" stage. The well-to-wheel analysis is commonly used to assess total energy consumption, or the energy conversion efficiency and emissions impact of marine vessels, aircraft and motor vehicles, including their carbon footprint, and the fuels used in each of these transport modes.

The well-to-wheel variant has a significant input on a model developed by the Argonne National Laboratory. The Greenhouse gases, Regulated Emissions, and Energy use in Transportation (GREET) model was developed to evaluate the impacts of new fuels and vehicle technologies. The model evaluates the impacts of fuel use using a well-to-wheel evaluation while a traditional cradle-to-grave approach is used to determine the impacts from the vehicle itself. The model reports energy use, greenhouse gas emissions, and six additional pollutants: volatile organic compounds (VOCs), carbon monoxide (CO), nitrogen oxide (NOx), particulate matter with size smaller than 10 micrometre (PM10), particulate matter with size smaller than 2.5 micrometre (PM2.5), and sulfur oxides (SOx).

Economic Input–output Life Cycle Assessment

Economic input–output LCA (EIOLCA) involves use of aggregate sector-level data on how much environmental impact can be attributed to each sector of the economy and how much each sector purchases from other sectors. Such analysis can account for long chains (for example, building an automobile requires energy, but producing energy requires vehicles, and building those vehicles requires energy, etc.), which somewhat alleviates the scoping problem of process LCA; however, EIOLCA relies on sector-level averages that may or may not be representative of the specific subset of the sector relevant to a particular product and therefore is not suitable for evaluating the environmental impacts of products. Additionally the translation of economic quantities into environmental impacts is not validated.

Ecologically Based LCA

While a conventional LCA uses many of the same approaches and strategies as an Eco-LCA, the latter considers a much broader range of ecological impacts. It was designed to provide a guide to wise management of human activities by understanding the direct and indirect impacts on ecological resources and surrounding ecosystems. Developed by Ohio State University Center for resilience, Eco-LCA is a methodology that quantitatively takes into account regulating and supporting services during the life cycle of economic goods and products. In this approach services are categorized in four main groups: supporting, regulating, provisioning and cultural services.

Exergy Based LCA

Exergy of a system is the maximum useful work possible during a process that brings the system into equilibrium with a heat reservoir. Wall clearly states the relation between exergy analysis and resource accounting. This intuition confirmed by DeWulf and Sciubba lead to Exergo-economic accounting and to methods specifically dedicated to LCA such as Exergetic material input per unit of service (EMIPS). The concept of material input per unit of service (MIPS) is quantified in terms of the second law of thermodynamics, allowing the calculation of both resource input and service output in exergy terms. This exergetic material input per unit of service (EMIPS) has been elaborated for transport technology. The service not only takes into account the total mass to be transported and the total distance, but also the mass per single transport and the delivery time. The applicability of the EMIPS methodology relates specifically to transport system. This model has been further improved by Trancossi who has introduced the friction term, which has not been considered by original EMIPS model, and the key distinction between exergy disruption by payload and by vehicle, focusing on the losses due to vehicle and more effective evaluation of the processes and produced an effective assessment of today transport vehicles. This model is referenced by Indian "Road less traveled" model, which has been developed for minimizing the impact of transports in urban environment.

Life Cycle Energy Analysis

Life cycle energy analysis (LCEA) is an approach in which all energy inputs to a product are accounted for, not only direct energy inputs during manufacture, but also all energy inputs needed to produce components, materials and services needed for the manufacturing process. An earlier term for the approach was *energy analysis*.

With LCEA, the *total life cycle energy input* is established.

Energy Production

It is recognized that much energy is lost in the production of energy commodities themselves, such as nuclear energy, photovoltaic electricity or high-quality petroleum products. *Net energy content* is the energy content of the product minus energy input used during extraction and conversion, directly or indirectly. A controversial early result of LCEA claimed that manufacturing solar cells requires more energy than can be recovered in using the solar cell. The result was refuted. Another new concept that flows from life cycle assessments is Energy Cannibalism. Energy Cannibalism refers to an effect where rapid growth of an entire energy-intensive industry creates a need for energy that uses (or cannibalizes) the energy of existing power plants. Thus during rapid growth the industry as a whole produces no energy because new energy is used to fuel the embodied energy of future power plants. Work has been undertaken in the UK to determine the life cycle energy (alongside full LCA) impacts of a number of renewable technologies.

Energy Recovery

If materials are incinerated during the disposal process, the energy released during burning can be harnessed and used for electricity production. This provides a low-impact energy source, especially when compared with coal and natural gas While incineration produces more greenhouse gas emissions than landfilling, the waste plants are well-fitted with filters to minimize this negative impact. A recent study comparing energy consumption and greenhouse gas emissions from landfilling (without energy recovery) against incineration (with energy recovery) found incineration to be superior in all cases except for when landfill gas is recovered for electricity production.

Criticism

A criticism of LCEA is that it attempts to eliminate monetary cost analysis, that is replace the currency by which economic decisions are made with an energy currency. It has also been argued that energy efficiency is only one consideration in deciding which alternative process to employ, and that it should not be elevated to the only criterion for determining environmental acceptability; for example, simple energy analysis does not take into account the renewability of energy flows or the toxicity of waste products; however the life cycle assessment does help companies become more familiar with environmental properties and improve their environmental system. Incorporating Dynamic LCAs of renewable energy technologies (using sensitivity analyses to project future improvements in renewable systems and their share of the power grid) may help mitigate this criticism.

In recent years, the literature on life cycle assessment of energy technology has begun to reflect the interactions between the current electrical grid and future energy technology. Some papers have focused on energy life cycle, while others have focused on carbon dioxide (CO_2) and other greenhouse gases. The essential critique given by these sources is that when considering energy technology, the growing nature of the power grid must be taken into consideration. If this is not done, a given class of energy technology may emit more CO_2 over its lifetime than it mitigates.

A problem the energy analysis method cannot resolve is that different energy forms (heat, electricity, chemical energy etc.) have different quality and value even in natural sciences, as a consequence of the two main laws of thermodynamics. A thermodynamic measure of the quality of energy is exergy. According to the first law of thermodynamics, all energy inputs should be accounted with equal weight, whereas by the second law diverse energy forms should be accounted by different values.

The conflict is resolved in one of these ways:

- value difference between energy inputs is ignored,

- a value ratio is arbitrarily assigned (e.g., a joule of electricity is 2.6 times more valuable than a joule of heat or fuel input),

- the analysis is supplemented by economic (monetary) cost analysis,

- exergy instead of energy can be the metric used for the life cycle analysis.

Critiques

Life cycle assessment is a powerful tool for analyzing commensurable aspects of quantifiable systems. Not every factor, however, can be reduced to a number and inserted into a model. Rigid system boundaries make accounting for changes in the system difficult. This is sometimes referred to as the boundary critique to systems thinking. The accuracy and availability of data can also contribute to inaccuracy. For instance, data from generic processes may be based on averages, unrepresentative sampling, or outdated results. Additionally, social implications of products are generally lacking in LCAs. Comparative life-cycle analysis is often used to determine a better process or product to use. However, because of aspects like differing system boundaries, different statistical information, different product uses, etc., these studies can easily be swayed in favor of one product or process over another in one study and the opposite in another study based on varying parameters and different available data. There are guidelines to help reduce such conflicts in results but the method still provides a lot of room for the researcher to decide what is important, how the product is typically manufactured, and how it is typically used.

An in-depth review of 13 LCA studies of wood and paper products found a lack of consistency in the methods and assumptions used to track carbon during the product lifecycle. A wide variety of methods and assumptions were used, leading to different and potentially contrary conclusions – particularly with regard to carbon sequestration and methane generation in landfills and with carbon accounting during forest growth and product use.

Streamline LCA

This process includes three steps. First, a proper method should be selected to combine adequate accuracy with acceptable cost burden in order to guide decision making. Actually, in LCA process, besides streamline LCA, Eco-screening and complete LCA are usually considered as well. However, the former one only could provide limited details and the latter one with more detailed information is more expensive. Second, single measure of stress should be selected. Typical LCA output includes resource consumption, energy consumption, water consumption, emission of CO_2, toxic residues and so on. One of these outputs is used as the main factor to measure in streamline LCA. Energy consumption and CO_2 emission are often regarded as "practical indicators". Last, stress selected in step 2 is used as standard to assess phase of life separately and identify the most damaging phase. For instance, for a family car, energy consumption could be used as the single stress factor to assess each phase of life. The result shows that the most energy intensive phase for a family car is the usage stage.

Life Cycle Assessment of Engineered Material in Service plays a significant role in saving energy, conserving resources and saving billions by preventing premature failure of critical engineered component in a machine or equipment.LCA data of surface engineered materials are used to improve life cycle of the engineered component.Life cycle improvement of industrial machineries and equipments including, manufacturing, power generation,transportaions etc leads to improvement in energy efficiency, sustainability and negating global temperature rise. Estimated reduction in anthropogenic carbon emission is minimum 10% of the global emission.

References

- Hawksworth, David L.; Bull, Alan T. (2008). Biodiversity and Conservation in Europe. Springer. p. 3390. ISBN 1402068646.

- Faber, M., Niemes, N. and Stephan, G. (2012). Entropy, environment, and resources, Springer Verlag, Berlin, Germany, ISBN 3642970494.

- Ruth, M. (1993). Integrating economics, ecology, and thermodynamics, Kluwer Academic Publishers, ISBN 0792323777.

- Oppenlander, Richard (2013). Food Choice and Sustainability. Minneapolis, MN: Langdon Street Press. pp. 120–123. ISBN 978-1-62652-435-4.

- Pearce, R. (2006). When the rivers run dry: Water – the defining crisis of the twenty-first century, Beacon Press, ISBN 0807085731.

- Smith, G. (2012). Nuclear roulette: The truth about the most dangerous energy source on earth, Chelsea Green Publishing, ISBN 160358434X.

- Miller GT (2004), Sustaining the Earth, 6th edition. Thompson Learning, Inc. Pacific Grove, California. Chapter 9, pp. 211–216, ISBN 0534400876.

- Oppenlander, Richard (2013). Food Choice and Sustainability. Minneapolis, MN: Langdon Street Press. p. 31. ISBN 978-1-62652-435-4.

- Galloway, J. N. (2003). "The Global Nitrogen Cycle", pp. 557–583 in H. D. Holland and K. K. Turekian (eds.) Treatise on Geochemistry. Pergamon Press, Oxford, ISBN 0080437516.

- Brinkman, Norman; Eberle, Ulrich; Formanski, Volker; Grebe, Uwe-Dieter; Matthe, Roland (15 April 2012). "Vehicle Electrification - Quo Vadis". VDI. Retrieved 27 April 2013.

Permissions

Index

www.ingramcontent.com/pod-product-compliance
Lightning Source LLC
Chambersburg PA
CBHW061930190326
41458CB00009B/2708